TECHNOLOGY AND INTERPRETING

Offering a thorough exploration of the rapid advancements in technology development for interpreting, this book explores the effects of technology on the practice of interpreting, providing theoretical insights and practical applications.

The chapters underscore the interdisciplinary nature of interpreting in the digital age and cover a wide range of topics from online learning environments for trainee interpreters and ethical considerations to the application of specific digital tools in language interpretation. It covers the application of advanced language models in interpreting, the pedagogical implications of technology in interpreter training as well as the cognitive aspects of interpreting in a technologically advanced environment. The authors integrate insights from technology, linguistics, education and cognitive science to provide a comprehensive understanding of how technology influences interpreting. This interdisciplinary approach allows the book to provide a holistic view of the field, making it a valuable resource for readers from various disciplines interested in the interplay between technology and interpreting. This book serves as a platform for discussing and debating new thoughts on the teaching and training of interpreting and the delivery of interpreting services.

Delving into the new paradigms of interpreting studies that have emerged in the wake of technological advancements, this book will be invaluable to students studying interpreting or related fields, researchers investigating the impact of technology on interpreting and practitioners in the field who are keen to understand and leverage the latest technological advancements.

Andrew K.F. Cheung is Associate Professor at the Hong Kong Polytechnic University. He is a member of the editorial boards of a number of journals. He is also a member of the Association International des Interprétes de Conference (AIIC). His research interests include computer-assisted interpreting and corpus-based interpreting studies.

Dechao Li is Professor at the Hong Kong Polytechnic University. He also serves as the chief editor of *Translation Quarterly*. His main research areas include corpus-based translation studies, empirical approaches to translation process research, history of translation and PBL and translator/interpreter training.

Kanglong Liu is Associate Professor at the Hong Kong Polytechnic University. His research interests include corpus-based translation studies, language and translation pedagogy and Hongloumeng translation. He is currently the associate editor of *Translation Quarterly*. He has published widely in scholarly journals and authored a monograph.

Riccardo Moratto is Distinguished Professor (特聘教授) in the School of Foreign Languages at Tongji University, Vice Director of the Research Center for Chinese Discourse and Global Communication, AIIC member, chartered linguist and fellow member of CIoL and general editor of two Routledge series.

TECHNOLOGY AND INTERPRETING

Navigating the Digital Age

Edited by Andrew K.F. Cheung, Dechao Li, Kanglong Liu and Riccardo Moratto

Routledge
Taylor & Francis Group

LONDON AND NEW YORK

First published 2026
by Routledge
4 Park Square, Milton Park, Abingdon, Oxon OX14 4RN

and by Routledge
605 Third Avenue, New York, NY 10158

Routledge is an imprint of the Taylor & Francis Group, an informa business

British Library Cataloguing-in-Publication Data
A catalogue record for this book is available from the British Library

Library of Congress Cataloging-in-Publication Data
Names: Cheung, Andrew K. F. editor | Li, Dechao, 1976– editor |
Liu, Kanglong editor | Moratto, Riccardo, 1985– editor
Title: Technology and interpreting: navigating the digital age / edited by Cheung Andrew K. F., Dechao Li, Kanglong Liu and Riccardo Moratto.
Description: Abingdon, Oxon; New York, NY: Routledge, 2026. |
Includes bibliographical references and index.
Identifiers: LCCN 2025006139 (print) | LCCN 2025006140 (ebook) |
ISBN 9781032982403 hardback | ISBN 9781032982373 paperback |
ISBN 9781003597711 ebook
Subjects: LCSH: Translating and interpreting—Technological innovations |
LCGFT: Essays
Classification: LCC P306.97.T73 T426 2026 (print) | LCC P306.97.T73 (ebook) |
DDC 418/.020285—dc23/eng/20250508
LC record available at https://lccn.loc.gov/2025006139
LC ebook record available at https://lccn.loc.gov/2025006140

ISBN: 9781032982403 (hbk)
ISBN: 9781032982373 (pbk)
ISBN: 9781003597711 (ebk)

DOI: 10.4324/9781003597711

Typeset in Times New Roman
by codeMantra

CONTENTS

MORE ABOUT THE EDITORS

Andrew K.F. Cheung is Associate Professor at the Hong Kong Polytechnic University. He holds a PhD from the University of East Anglia. He is a member of the editorial boards of a number of journals. He is also a member of the Association International des Interprétes de Conference (AIIC). His research interests include computer-assisted interpreting and corpus-based interpreting studies.

Dechao Li is Professor in the Department of Chinese and Bilingual Studies, The Hong Kong Polytechnic University. He also serves as the chief editor of *Translation Quarterly*, a journal published by the Hong Kong Translation Society. His main research areas include corpus-based translation studies, empirical approaches to translation process research, history of translation in the late Qing and early Republican periods and PBL and translator/interpreter training.

Kanglong Liu is Associate Professor at The Hong Kong Polytechnic University. His research interests include corpus-based translation studies, language and translation pedagogy and Hongloumeng translation research. He is currently the associate editor of *Translation Quarterly*, the official publication of the Hong Kong Translation Society. He has published widely in scholarly journals and authored the monograph *Corpus-Assisted Translation Teaching: Issues and Challenges* (2020).

Riccardo Moratto is Distinguished Professor (特聘教授) in the School of Foreign Studies (SFS), Tongji University, Deputy Director of the Research Center for Chinese Discourse and Global Communication, AIIC member, expert member of the Translators Association of China (TAC), Standing Council Member of World Interpreter and Translator Training Association (WITTA), and member of numerous other associations. Professor Moratto is the executive editor of the *International*

Journal of Translation and Communication, editor-in-chief of *Interpreting Studies for Shanghai Foreign Language Education Press* (外教社) and general editor of *Routledge Studies in East Asian Interpreting* and *Routledge Interdisciplinary and Transcultural Approaches to Chinese Literature.* Professor Moratto is a chartered linguist and fellow of the Chartered Institute of Linguists (FCIL), a hyperpolyglot, an international conference interpreter and a renowned literary translator. Professor Moratto has published extensively in the fields of translation and interpreting studies and Chinese literature. Prof. Moratto would like to acknowledge the following grant: 中央高校基本科研业务费专项资金资助 (Supported by the Fundamental Research Funds for the Central Universities).

CONTRIBUTORS

Vincent Chieh-Ying Chang is currently Assistant Professor in the Department of English, Tamkang University, Taiwan. He holds a BA in English from National Chengchi University, Taiwan and later became a professional conference interpreter in 1999 upon completion of a two-year post-graduate training on translation and conference interpreting at Fujen University, Taiwan. While holding an MSc in Media and Communications from LSE (London School of Economics), he completed his PhD in Translation Studies from Imperial College London. His research interests lie in the cognitive aspects of T&I, multimodality and religious T&I.

Po-Lin Chen holds a PhD in Education from National Chengchi University, Taiwan. He is currently Professor in the Department of Psychology and Counseling at the National Taipei University of Education, Taipei, Taiwan. He is committed to teaching courses in psychological testing and positive psychology. His research interests include positive psychology, educational psychology and psychological testing.

Andrew K.F. Cheung is Associate Professor at the Hong Kong Polytechnic University. He holds a PhD from the University of East Anglia. He is a member of the editorial boards of a number of journals. He is also a member of the Association International des Interprétes de Conference (AIIC). His research interests include computer-assisted interpreting and corpus-based interpreting studies.

Feng Cui is a senior lecturer and PhD supervisor in the Chinese Programme at Nanyang Technological University (NTU), Singapore. He also serves as the coordinators of the Minor in Translation program and the Han Suyin Scholarship Fund (in Translation Studies) at NTU. Dr. Cui's research interests include translation history in China, translation theories, 20th-century Chinese literature and comparative literature. He has authored, co-authored, edited and translated eight books and published over 50 journal articles and book chapters. His work appears in prestigious journals indexed in SSCI, A&HCI, CSSCI and THCI. Dr. Cui is currently the editor of *Humanities and Social Sciences Communications* (SSCI and A&HCI) and guest editor of a Special Issue titled "Transforming Translation Education through Artificial Intelligence" in *The Interpreter and Translator Trainer* (SSCI and A&HCI).

Fei Deng is Full Professor in the School of Foreign Studies, South China Agricultural University. Her research includes corpus linguistics, EFL (English as a foreign language) / ESL (English as a second language) curriculum and teaching methodology, corpus-assisted discourse analysis and translation studies.

Damien Chiaming Fan is Associate Professor in the Graduate Program in Translation and Interpretation at National Taiwan University. He received his interpreter training and PhD from National Taiwan Normal University. In addition to the pedagogy and cognitive aspects of conference interpreting, he is interested in its paradigm shift in the post-COVID world.

Wei Guo is Associate Professor in the School of Foreign Languages, Central South University, where she also serves as a Master's Supervisor and Director of the MTI Education Center. A Fulbright Foreign Language Teaching Assistant in both China and the United States, her research focuses on interpreting pedagogy, translation studies and the integration of technology in language education. She has published over 30 academic papers, including nine in SSCI and CSSCI journals, and has led several major research projects, including a Humanities and Social Sciences Youth Project funded by the Ministry of Education of China.

Dan Feng Huang is a doctoral student at Hong Kong Polytechnic University and an in-service teacher at the Guangdong Polytechnic Normal University. Her research interests include corpus-based linguistics, psycholinguistics and formulaic language. She is particularly interested in the psychological realities involved in using formulaic sequences in the context of bilingualism.

Zi-ying Lee is Assistant Professor at the National Changhua University of Education in Taiwan. Her research interests include expert demonstration and interpreter training, reflective practice in interpreting, student self-assessment, interpreting practicum, translation pedagogy and retranslation studies.

Dechao Li is Professor in the Department of Chinese and Bilingual Studies, The Hong Kong Polytechnic University. He also serves as the chief editor of *Translation*

Quarterly, a journal published by the Hong Kong Translation Society. His main research areas include corpus-based translation studies, empirical approaches to translation process research, history of translation in the late Qing and early Republican periods and PBL (Problem based learning) and translator/interpreter training.

Heming Li is a graduate student of English Interpreting in the School of Foreign Languages, China University of Petroleum (East China). Her current research interests include interpreting ethics, aptitude for interpreting and interpreting cognition.

Zhi Li is Associate Professor in the School of Western Languages at Harbin Normal University, Heilongjiang Province. She works to promote translation technology education and is in charge of MTI translation technology courses in her department. Her research interests include interpreting and translation theories and practice, as well as interpreting and translation technologies. She has published more than ten articles in translation journals, including seven in core journals. She is the co-author of two books on interpreting and technologies, one of the editors for two translation textbooks and the leader of four research projects relating to interpreting and technology.

Yi-Ti Lin holds a PhD in English from Tamkang University, Taiwan. She is currently Professor in the Department of English at Tamkang University. She is dedicated to teaching English to EFL learners in Taiwan. Her research interests focus on second language acquisition, sociolinguistics and cross-cultural communication.

Kayo Matsushita is Professor in the College of Intercultural Communication and the Graduate School of Intercultural Communication at Rikkyo University. Drawing on her experiences as a conference interpreter and journalist, her main area of focus is translation and interpreting in the media, but she is also interested in issues surrounding technology and interpreting. She obtained her MS from the Graduate School of Journalism at Columbia University and her PhD from the Graduate School of Intercultural Communication at Rikkyo University. She is the author of *When News Travels East: Translation Practices by Japanese Newspapers* (2019).

David B. Sawyer worked as a conference interpreter and translator for over three decades, including over ten years as the principal diplomatic interpreter for German at the U.S. State Department. He served on the faculty at three universities and directed the program at the University of Maryland, College Park. He holds a Master of Education from the University of Illinois at Chicago and graduate degrees in translation and conference interpreting as well as a doctorate in interpreting studies from the University of Mainz. He currently serves as Director of Language Testing at the U.S. State Department's Foreign Service Institute.

Shuxian Song is a senior lecturer in the School of Foreign Studies, China University of Petroleum (East China) and holds a PhD in Translation and Interpreting at the Hong Kong Polytechnic University. Her research interests are corpus interpreting studies and cognitive translation studies. She has published relevant articles in several influential journals.

Kan Wu is Associate Professor of Translation and Interpreting Studies in Dongfang College, Zhejiang University of Finance and Economics. He also holds a postdoctoral fellowship in Translation and Interpreting Studies at the University of Macau. His research primarily focuses on corpus-based translation and interpreting studies, as well as digital humanities.

Cui Xu is Assistant Professor in the School of Foreign Languages, Beijing Institute of Technology (BIT). She received her PhD degree from the Hong Kong Polytechnic University. She specializes in corpus-based translation and interpreting studies, and her main interests include empirical approaches to translation studies, corpus-based interpreting studies and language variation. She has published relevant research in several journals, including *Target, Babel, Linguistica Antverpiensia, Translation Quarterly* and *Foreign Language Teaching and Research*.

Masaru Yamada is Professor in the College and the Graduate School of Intercultural Communication at Rikkyo University. His current research focuses on translation processes, translation technologies (including CAT, MTPE and LLMs) and Translation in Language Teaching (TILT). He co-edited *Metalanguages for Dissecting Translation Processes: Theoretical Development and Practical Applications* (Routledge, 2022) and the special Ampersand issue *Empirical Translation Process Research* (2024). Other recent publications include "Optimizing Machine Translation through Prompt Engineering: An Investigation into ChatGPT's Customizability" (2023).

Pan Zhao is a doctoral candidate and student researcher at the Faculty of Humanities, The Hong Kong Polytechnic University. With a multidisciplinary background, he investigates the intersections of technology-mediated interpreting, translation and cross-cultural communication. His recent publications include book reviews on misinformation and online education.

INTRODUCTION

Andrew K. F. Cheung, Dechao Li, Kanglong Liu and Riccardo Moratto

This volume, *Technology and Interpreting: Navigating the Digital Age*, explores the interplay between interpreting and technology, offering insights into how digital tools are transforming the profession. The COVID-19 pandemic further accelerated the adoption of technology in interpreting, forcing interpreters and educators to reimagine traditional modes of delivery.

The use of remote simultaneous interpreting (RSI) has implications for both interpreters and listeners, and we do not yet know enough about these effects. The rise of RSI has benefits for both interpreters and conference attendees, as they no longer need to travel. Interpreters can work from anywhere, and attendees can tune in to follow events and listen to simultaneous interpreting from any location (Cheung 2024). RSI allows interpreters to provide their services beyond geographical limitations, but they may have to use a language variant they do not habitually use. This can impact the accuracy of RSI, as interpreters might face increased cognitive load and mental fatigue. Additionally, RSI listeners often miss out on para-linguistic and extra-linguistic cues present in onsite interpreting, which can affect their perception of interpreting quality (Cheung 2022). This lack of para-linguistic and extra-linguistic cues in RSI environments can lead to a greater focus on formal elements such as the interpreter's accent, fluency and delivery, particularly when listeners do not have access to the source language. This reveals how technology reshapes perceptions of interpreting quality, making it essential for interpreters to adapt to these new contexts.

Moreover, automatic speech recognition (ASR)-based computer-assisted interpreting tools have emerged as a game-changing innovation, particularly for increasing accuracy in rendering challenging elements like numbers and technical terms (Defrancq et al. 2024). Yet, their integration into the booth introduces complexities, including increased cognitive load and disruptions in fluency, as

DOI: 10.4324/9781003597711-1

interpreters must balance auditory input, visual information and decision-making processes (Li and Chmiel 2024). The adoption of interpreting technologies also necessitates new skills and competencies (Mellinger 2023). Interpreter training programs must evolve to include technology-focused curricula that prepare students for technology-mediated environments. Additionally, the evolution of interpreting technologies is not unidirectional; interpreters themselves drive innovation by articulating their needs and challenges (Pöchhacker and Liu 2024).

The rapid development of technologies further accelerates the adoption of digital tools in interpreting, forcing interpreters and educators to reimagine traditional modes of delivery. Online learning environments, distance interpreting (DI) and hybrid models have become the new norm, requiring interpreters to develop new skills and strategies. At the same time, emerging technologies such as generative artificial intelligence (AI), virtual reality (VR) and electronic corpora are opening new frontiers for interpreting education and practice. The contributions in this volume explore these dynamic changes and provide a roadmap for navigating the digital age.

Chapter summaries

In Chapter 1, Vincent Chieh-Ying Chang, Yi-Ti Lin and Po-Lin Chen investigate the effects of tech-driven online learning environments during the COVID-19 pandemic on concentration and metacognitive learning strategies for interpreter trainees in the greater Chinese-speaking region. Their study reveals a positive interplay between online learning environments, learner concentration and metacognitive strategies, highlighting the potential of online platforms to enhance interpreter training and flexibility during challenging times.

In Chapter 2, Damien Chiaming Fan reflects on the evolution of DI in Taiwan during and after the pandemic. By combining surveys and interviews, the chapter explores changes in workload, working conditions and attitudes toward DI. While DI offers benefits such as increased flexibility and reduced commuting, it poses challenges like technical issues and lack of personal interaction. The chapter also highlights the emergence of hybrid interpreting and the growing need for interpreters to balance technological proficiency with core interpreting skills.

In Chapter 3, Wei Guo and Feng Cui conduct a systematic review of technology integration in interpreter education, focusing on current applications, challenges and regional variations. Using Preferred Reporting Items for Systematic Reviews and Meta-Analyses guidelines, the authors examine computer-assisted interpreter training tools and their implications for students, instructors and institutions. The chapter emphasizes the importance of fostering technological innovation while addressing regional differences in adopting such tools.

In Chapter 4, Dan Feng Huang and Fei Deng explore the use of corpus tools, Sketch Engine and AntConc, to compare lexical bundles in interpreted and native

Chinese. Their analysis identifies unique linguistic patterns in interpreted Chinese, such as explicitation and preferences for certain lexical structures, demonstrating the utility of corpus technology in interpreting research and practice.

In Chapter 5, Zhi Li examines the impact of technology on interpreting education, focusing on tools such as ASR, cloud technology and AI-generated content. The chapter provides insights into how these technologies influence teaching methods, including VR labs and digital resources, and offers recommendations for interpreter training institutions to adapt to these rapid changes.

In Chapter 6, David B. Sawyer discusses the implications of AI and machine learning (ML) for interpreting education. The chapter highlights how neural machine translation and AI-driven tools are reshaping the curriculum, urging educators to rethink traditional models and incorporate AI/ML competencies to prepare students for the challenges of the AI era.

In Chapter 7, Shuxian Song and Heming Li delve into the ethical considerations of using technology in interpreting. The chapter addresses issues such as confidentiality, data security and intellectual property rights, emphasizing the importance of balancing human-machine collaboration with ethical standards to ensure responsible integration of technology in the interpreting profession.

In Chapter 8, Kan Wu and Dechao Li explore the integration of VR and augmented reality in interpreter training. The chapter highlights the potential of immersive technologies to enhance engagement, simulate complex scenarios and provide adaptive learning experiences. It also addresses the challenges of implementing these technologies, including financial constraints and ethical concerns, while proposing effective methodologies for their integration.

In Chapter 9, Cui Xu reviews the role of electronic corpora in interpreter training and education. The chapter underscores the growing importance of corpus linguistics in interpreting studies, discussing its practical applications and the challenges of compiling interpreting-specific corpora. The chapter calls for greater integration of corpora into interpreter training programs to enhance linguistic analysis and educational outcomes.

In Chapter 10, Masaru Yamada and Kayo Matsushita investigate the potential of Large Language Models (LLMs), such as ChatGPT-4, in assessing interpreting quality. Their study explores innovative prompts like zero-shot and chain-of-thought strategies to evaluate simultaneous interpreting. The chapter highlights the potential of LLMs to provide scalable and consistent feedback while noting the limitations of text-based evaluations in capturing nuanced interpreting performance.

In Chapter 11, Pan Zhao and Andrew K.F. Cheung examine the professional ethics of institutional interpreters in Mainland China through a case study of live simultaneous interpreting at the 2022 Beijing Paralympic Games. The chapter uncovers tensions between institutional and interpreting ethics, revealing how political awareness and national interests influence interpreters' decision-making in institutional contexts.

In Chapter 12, Zi-ying Lee investigates the use of speech-to-text (STT) technology and expert demonstrations in interpreter training. The chapter demonstrates how these tools support self-assessment, enabling students to identify delivery issues and improve interpreting strategies. The findings highlight the potential of STT to enhance deliberate practice and foster independent learning among interpreter trainees.

Conclusion

The chapters in this volume collectively illustrate the profound impact of technology on interpreting education, practice and ethics. From online learning environments and DI to AI-driven tools and VR, the contributions highlight the opportunities and challenges of navigating the digital age. By addressing these developments from multiple perspectives, technology and interpreting: navigating the digital age offers a comprehensive resource for interpreters, educators and researchers seeking to adapt to the rapidly evolving landscape of the interpreting profession. This volume not only sheds light on the current state of the field but also provides a foundation for future research and innovation in interpreting in the digital era.

References

Cheung, Andrew K. F. 2022. "Listeners' Perception of the Quality of Simultaneous Interpreting and Perceived Dependence on Simultaneous Interpreting." *Interpreting* 24(1): 38–58. https://doi.org/10.1075/intp.00070.che.

Cheung, Andrew K. F. 2024. "Cognitive Load in Remote Simultaneous Interpreting: Place Name Translation in Two Mandarin Variants." *Humanities and Social Sciences Communications* 11(1): 1238. https://doi.org/10.1057/s41599-024-03767-y.

Defrancq, Bart, Helena Snoeck and Claudio Fantinuoli. (2024). "Interpreters' performances and cognitive load in the context of a CAI Tool." In Marion Winters, Sharon Deane-Cox & Ursula Böser (Eds.), *Translation, Interpreting and Technological Change: Innovations in Research, Practice and Training*. London: Bloomsbury Academic, 37–58.

Li, Tianyun, and Agnieszka Chmiel. 2024. "Automatic Subtitles Increase Accuracy and Decrease Cognitive Load in Simultaneous Interpreting." *Interpreting* 26(2): 253–81. https://doi.org/10.1075/intp.00111.li.

Mellinger, Christopher .D. (2023). "Embedding, extending, and distributing interpreter cognition with technology." In G. Corpas Pastor & B. Defrancq (Eds.), *Interpreting technologies — current and future trends*. Amsterdam: John Benjamins, 195–216.

Pöchhacker, Franz, and Minhua Liu. 2024 "Interpreting Technologized: Distance and Assistance." *Interpreting* 26(2): 157–77. https://doi.org/10.1075/intp.00112.poc.

1

TECH-ENABLED FOCUS AND FLOURISH

Navigating online learning environments, concentration and metacognitive strategies for trainee interpreters

Vincent Chieh-Ying Chang, Yi-Ti Lin and Po-Lin Chen

1.1 Introduction

1.1.1 Translator and interpreter training in higher education in the greater Chinese-speaking region

In the past two decades, there has been a substantial increase in the number of postgraduate translation and interpreting (T&I) programs throughout the Chinese-speaking region (Setton 2011; Tao 2016; Zhong 2017). A number of universities in this region have also begun to offer tertiary T&I training courses[1] as optional subjects for students enrolled in English-major programs at an undergraduate level (cf. Cai and Dong 2015; Kumar 2017; Zhan 2014).

This trend in higher education (HE) is evidence of a growing recognition of the importance of language skills in the greater Chinese-speaking region (Kramsch 2014). There are several reasons for this (Wang and Lei 2009). First, as the world economy has become increasingly globalized, successful corporations need individuals with excellent inter-lingual and cross-cultural communication skills (see Biel and Sosoni 2017; Lin 2015; Liu et al. 2024). Second, students themselves desire to improve their foreign language proficiency for social reasons (Pym and Ayvazyan 2017; Pym et al. 2013). Finally, these courses equip students for careers as translators and interpreters (Lee and Liao 2010).

1.1.2 Tertiary trainee interpreters as an under-researched population

Research in the field of T&I studies has seen a dramatic surge over the past two decades (cf. Yan et al. 2017), including a wide range of topics related to T&I programs

DOI: 10.4324/9781003597711-2

in HE (Kelly and Martin 2019). The majority of these academic endeavors involve T&I "postgraduate students" or "professional trainees" (Yan et al. 2017, 6) as participants and study a variety of T&I issues classified within the various "maps" of T&I studies (for examples of different maps, see Chesterman 2009; Holmes 2021; Toury 1995; Van Doorslaer 2007; Williams and Chesterman 2002; Zanettin et al. 2015).

However, this research has long overlooked undergraduate or tertiary T&I learners. In particular, little is known about the metacognitive learning strategies they employ and the typical difficulties they encounter (Chang and Chen 2023; Yan et al. 2017, 5). There are at least two reasons for this neglect. One is that tertiary interpreter trainees are often a hidden population within university communities as they are not typically enrolled in conventional academic programs (Xu 2005). Additionally, unlike postgraduate MA programs, tertiary interpreter training is often seen as a mere add-on component to undergraduate foreign language degree programs that primarily aim to help their students master the language and literature of a foreign culture, rather than T&I skills specifically (Pym and Ayvazyan 2017; Pym et al. 2013).

Tertiary trainee interpreters are, therefore, a significantly under-researched population. A number of scholars (e.g., Laviosa 2014; Leonardi 2010; Malmkjær 2010) have suggested that this represents a serious omission, given the growing global importance of T&I skills and the obvious popularity of one-year T&I modules among undergraduate language majors (Liu 2020, 43).

1.1.3 Drastic expansion of online learning in HE and for interpreter training amidst the COVID-19 pandemic

Online learning, where synchronous and asynchronous learning modes can be applied together (Graham 2013; Graham et al. 2013), has emerged as a potential solution to the pedagogical problems posed by the pandemic (Mahaye 2020; Singh 2021). It has been shown to be effective in many different disciplines (e.g., Balakrishnan et al. 2021; Marita and Utami 2020; Sitthiworachart et al. 2021; Wuwung and Tulung 2021). In the field of T&I studies, in particular, it has been successfully used in a great number of contexts (e.g., Bergunde and Pollabauer 2019; Eser et al. 2020; Martins et al. 2015; Melchor 2018; Moser-Mercer et al. 2005; Şahin 2013).

The pandemic has played havoc with university schedules around the world (Byrnes et al. 2021; Sohrabi et al. 2021), including those of T&I programs (Mok et al. 2021; Viner et al. 2020). In Chinese Mainland, for example, all universities switched to online delivery of classes in late January 2020 in an attempt to contain the spread of the virus (Kelvin and Rubino 2020; Nkengasong 2020). However, this sudden move to online learning has posed a number of challenges for T&I programs, which typically rely heavily on face-to-face interaction between students and instructors (Almahasees and Qassem 2021; Alwazna 2021; Perez and Hodakova 2021).

Further, the pandemic has resulted in an explosive increase in the use of information technology for learning, as conventional classroom teaching has been disrupted. This has led to an intense interest in the potential impact of such technology on learning outcomes involving a number of educational psychology constructs (cf. Halverson and Graham 2019; McInerney 2013). In particular, there is a need to better understand a number of issues:

1 Despite the social isolation imposed by the pandemic restrictions, are students capable of managing their learning environments in association with online learning?
2 Are the mental concentration levels of students sustained effectively during online teaching?
3 Are students able to adopt effective metacognitive learning strategies during their online learning experiences?

Online teaching has been proven to be a popular approach to learning that combines traditional classroom learning with online platforms and resources. While the use of technology in the learning environment has many benefits, such as improving concentration and facilitating access to information, it can also have certain drawbacks. In particular, recent research has suggested that factors such as social isolation (Lane et al. 2021; Mozelius 2020), changes in concentration levels (Arkhipova et al. 2018; Yang et al. 2021), and metacognitive learning strategies may be adversely affected (Broadbent 2017; Resien et al. 2020) by the use of technology in the learning environment.

As noted by Lane et al. (2021) in an article published on how swiftly most Chinese universities moved to online delivery (see also Jiang et al. 2021), the pandemic has had major repercussions on institutions of HE around the world (Crawford et al. 2020; Pokhrel and Chhetri 2021). For example, many American universities have suspended their spring 2020 semesters or at least changed them to remote learning environments (Crawford et al. 2020; Guangul et al. 2020). Additional instances in Asia are all universities in Hong Kong that changed their regular learning environment to a remote situation (Moorhouse 2020; Xiao and Li 2020). Universities in other parts of the globe have gone even further, requiring all classes to move completely online (e.g., Al-Baadani and Abbas 2020; Drane et al. 2020; Haider and Al-Salman 2020; Toquero 2020).

1.2 Social constructivist approach to tertiary interpreter training using online learning

In the field of T&I studies, Kiraly (2014) has proposed a social constructivist theoretical approach, drawing on prior pioneering research (Brown et al. 1989; Bruffee 1999; Dewey 1986; Glasersfeld 1988; Rorty 2009; Vygotsky 1994), to examine how T&I learning takes place. Social constructivists believe that T&I competence is developed through personal interactions with others, rather than being simply "given" or "acquired" (Kiraly 2014, 62).

The approach has been applied by several researchers to evaluate the success of online learning for professional interpreter competence (e.g., Galán-Mañas and Albir 2010; Pan and Yan 2012; Pokorn 2009; Presas 2012). It is particularly relevant because it takes into account the key influence of the social and cultural context in which learning takes place (Sandrelli 2002; Sandrelli and Jerez 2007) and thus helps to identify areas where online learning may be more or less effective (Eser et al. 2020; Hansen and Shlesinger 2007; Ko 2006) or can be improved (Class et al. 2004; Moser-Mercer 2008; Mouzourakis 2006).

It is important to note that social constructivism is not a single coherent theory, but an umbrella term for a range of ideas and approaches. Nevertheless, it does provide a comprehensive framework for understanding the potential interplay(s) of factors that influence the success of interpreter training during online learning. Kiraly (2014) has explicitly identified three factors, in particular, which are the focus of this current study: the student's online learning environment, their mental concentration and their preferred metacognitive learning strategies.

1.2.1 *Online learning environment*

To create a successful online learning environment for interpreter training, it is necessary to focus on three areas: the physical environment, the social environment and the cognitive environment (Kiraly 2014, 81–82). The physical environment includes factors such as the design of the learning space, learning equipment and quietness level (González-Davies and Enríquez-Raído 2016; Varney 2009). The social environment includes the social interaction between learners and instructors, as well as the social norms that are established within the learning community (Kiraly 2014, 2015). The cognitive environment includes the way information is presented, sequenced and reviewed (Chang 2011; Chang et al. 2008; Massey 2005; Thomas et al. 2014).

Furthermore, online learning depends on the proficient use of technological devices, stable access to the internet and digital literacy of learners (Adedoyin and Soykan 2020). Therefore, the viability of the learning environment and the learner's ability to solve simple technical problems are critical for successful and efficient online learning (Mishra et al. 2020). In addition to ensuring that all learners are equipped with appropriate mobile devices and are provided with appropriate technology infrastructure and internet access, skills and awareness of the learners to create an appropriate online learning environment are also important for effective T&I training (Mishra et al. 2020).

1.2.2 *Concentration*

There is no doubt that concentration is a key factor in effective learning, whether online or offline (Ismail et al. 2016). In particular, concentration is indispensable to effective online learning for tertiary interpreter training during the COVID-19 pandemic (Adedoyin and Soykan 2020).

Additionally, the asynchronous nature of online learning means that learners need to be able to focus and concentrate for extended periods of time in order to absorb and process the materials (Singh 2021). This can be challenging for learners, especially those who are used to the more interactive and immediate nature of offline learning (Graham et al. 2013).

Further, concentration is even more important in online learning environments, where learners need to be able to switch between synchronous and asynchronous learning modes (Balakrishnan et al. 2021). In order to effectively learn in an online environment, learners need to be able to concentrate in both synchronous and asynchronous settings (Halverson and Graham 2019).

To maintain focus and motivation during online learning for interpreter training, it is important to avoid distractions and interruptions as much as possible. This can be facilitated through creating a designated workspace (Kiraly and Hofmann 2016; Risku 2016) and minimizing social media use (Desjardins 2011; Robinson et al. 2016).

1.2.3 Metacognitive learning strategies

Metacognitive learning strategies are particularly important in online learning, as they help learners to keep track of their learning and progress (Motta 2016). Additionally, these strategies can help learners to identify areas where they need more support (Ismail et al. 2016).

Some metacognitive learning strategies that are particularly effective for online learning include setting learning goals, monitoring progress and seeking feedback (Resien et al. 2020). By using these strategies, learners can make sure that they are successfully progressing in their learning, even when they are not in a traditional classroom setting (Resien et al. 2020).

Various metacognitive learning strategies that can be adopted, either singly or in combination, by students during online learning for interpreter competence. These include cognitive apprenticeship (Kiraly 2014, 47–49), repeated review of teaching materials (Motta 2016; Thomas et al. 2014) and peer-based learning (Kiraly 2014, 111–15).

1.2.4 Motivation, significance and contribution

1.2.4.1 Motivation

The outbreak of COVID-19 in the greater Chinese-speaking region has seriously affected university teaching and learning activities (Mok et al. 2021). To minimize the impact of the pandemic, many universities have switched to online (synchronous and asynchronous combined) teaching and learning modalities. However, little is known about how online learning affects tertiary interpreter training. This has therefore motivated the present study to design an investigation to explore the potential effect(s) of online learning on three primary learning factors: the online learning environment, concentration level and metacognitive learning strategies (Kiraly 2014).

1.2.4.2 Significance

Echoing Pintrich (1991), Kiraly (2014) has pointed out that the online learning environment, concentration and metacognitive learning strategies are three important educational psychology constructs that have been shown to impact interpreter training. Nonetheless, the potential interplay between these three constructs in the context of interpreter training has received limited research attention. The COVID pandemic presents an inimitable opportunity to examine the interrelationship between these three constructs during tertiary online interpreter training. This pandemic backdrop has therefore yielded such a research-worthy problem of great significance that warrants academic investigations.

1.2.4.3 Contribution

Further, the COVID pandemic presents a rare ontology where researchers can investigate the associations, if any, between the online learning environment, concentration and metacognitive learning strategies accompanying tertiary online interpreter training. As a novel contribution to knowledge, the present study is the first attempt to explore the interrelationships, if any, between the said three factors. This research is extremely important and can generate significant social impact in the current context. With the COVID pandemic continuing to disrupt face-to-face interpreter training, it is essential that we understand how best to adapt and deliver T&I training online. This research will make a significant contribution to advancing our understanding of how to potentially create an effective online learning environment for tertiary interpreter trainees.

Additionally, as yet another potential societal impact, this research may also provide valuable insights into how to optimize tertiary online interpreter training in the context of the COVID pandemic, if not future acts of God.

1.2.5 Research questions

The present study is theoretically conceptualized, using the social constructivist approach (Kiraly 2014). Grounded within the social constructivist framework (Kiraly 2014), two research questions are therefore framed and raised as follows:

1 What is/are the effect(s), if any, of the online learning modality on the three factors identified above accompanying tertiary interpreter training during the COVID pandemic?

2 What associations, if any, occurred between these three factors in tertiary online interpreter training during the COVID pandemic?

Assuming associations, if at all, between the three factors were present, we proposed three hypotheses as follows:

a Hypothesis 1: Online learning environment is significantly associated with concentration during interpreter training.
b Hypothesis 2: Online learning environment is significantly associated with metacognitive strategies during interpreter training.
c Hypothesis 3: Concentration is significantly associated with metacognitive strategies during interpreter training.

1.3 Methods

1.3.1 Participants

A total of 291 undergraduates who took tertiary interpreter training as an (optional) course in the greater Chinese-speaking region were successfully recruited; 230 females and 61 males in the age range between 19 and 24 years old voluntarily agreed to participate with informed consent. All were English language majors who had previously gained some basic experience in interpreting but had received no formal postgraduate interpreting training prior to the investigation. They were all Chinese native speakers and learned English as a foreign language from five different societies in the greater Chinese-speaking region: Chinese Mainland, Hong Kong, Macau, Singapore and Taiwan (arranged alphabetically). This led to a total of 291 valid questionnaires used for data analyses.

1.3.2 Instruments

A questionnaire (Appendix 1) consisting of three dimensions, i.e., (1) online learning environment; (2) concentration; and (3) metacognitive learning strategies, was administered to the participants, in combination with five additional questions associated with the participants' demographics. All of the items were measured by a four-point Likert scale, ranging from 1 as strongly disagree to 4 as strongly agree. "Online learning environment" included three items on how learners create an appropriate online learning environment (adapted from Shim and Lee 2020). For example, "During distance learning for interpreter training, if there is a sudden problem with the software or the internet, I will try to solve the problem or seek help". The dimension of "concentration" consisted of three items on how learners concentrate during T&I class while learning online (adapted from Pintrich 1991). For example, "During synchronous distance learning (i.e., the teacher and the students are online at the same time) in the interpreting class, even if the teacher does not do a roll-call or does not ask us to turn on the camera to answer questions, I will still try to stay focused". The dimension on "metacognitive learning strategies" included two items on how the learners control their learning process (adapted from Pintrich 1991). For example, "During asynchronous distance learning in the interpreting class (i.e., teachers pre-recorded videos), I tend to pause at important passages and continue watching only after taking down notes or only after having gained a full understanding".

1.4 Results

1.4.1 Descriptive and correlation analysis

1.4.1.1 Descriptive results

Of the participants, approximately 79% were female and the other 21% were male, with an average age of 22. The participants had spent an average of 13 years studying English as a foreign language, with an average of three hours spent per week on their interpreting training at universities located in such cities as Beijing, Changsha, Chengdu, Guangzhou, Hangzhou, Hong Kong, Hsinchu, Kaohsiung, Macau, Shanghai, Shenzhen, Singapore, Taichung, Tainan, Taipei, Taiyuan, Xiamen, Xi'an, Zhanjiang and Zhuhai (arranged alphabetically).

1.4.1.2 Correlation analysis

Results yielded by a correlation analysis have shown that the three factors, i.e., (1) online learning environment; (2) concentration; and (3) metacognitive learning strategies were all positively related. The means, standard deviations and correlation coefficients of online learning environment, concentration and metacognitive learning strategies are presented in Table 1.1.

The mean scores of the three factors were 3.12 for online learning environment, 3.16 for concentration and 3.13 for metacognitive strategies, suggesting that the learners agreed on the importance of online learning environment, concentration and metacognitive strategies during online interpreter training. The achieved means for the three factors were similar, indicating that the learners had a consensus on the equal value of the factors.

TABLE 1.1 Correlation analysis

Measures	M	SD	1	2	3
Online learning environment	3.12	.39	1		
Concentration	3.16	.44	.68**	1	
Metacognitive strategies	3.13	.46	.52**	.36**	1

Notes: $N = 291$; **$p < .01$.

In answering the first research question, this means that the online learning mode had an effect on the three factors and that, to be successful in interpreter training, learners must set up a viable online learning environment for themselves, concentrate during class and make good use of metacognitive strategies to help them learn better. The standard deviation value of each factor ranges from .39 to .46, showing a moderate variance. The correlation among the three factors was medium to high, with $r = .68$ between online learning environment and concentration, $r = .36$ between concentration and metacognitive strategies and $r = .52$ between online learning environment and metacognitive strategies. The coefficients of the three factors show that the three factors were positively and significantly correlated with each other.

According to the correlational analysis of the three factors (Table 1.2), online learning environment and concentration showed a stronger correlation (ranging from $r = .36$ to $r = .63$), followed by concentration and metacognitive strategies (ranging from $r = .22$ to $r = .57$), and then, online learning environment and metacognitive strategies (ranging from $r = .42$ to $r = .52$). All the correlation coefficients were positive and significant, indicating the strong associations of the three factors.

1.4.2   Analyses of the measurement model and the structural model (for a diagrammatic view, see Figure 1.1)

The measurement model was assessed by confirmatory factor analysis, testing convergent validity, internal consistency and discriminant validity (Gerbing and Anderson 1988). The measurement model was tested first, followed by the testing of the structural modeling.

In T&I studies, it is worth emphasizing that this type of testing aims to model the correlational structure of different factors and is a novel way for analysis in the T&I discipline (Chang 2024).

TABLE 1.2 Means, standard deviations and zero-order correlational matrix ($N = 291$)

	M	SD	1	2	3	4	5	6	7	8
OLE1	3.18	.44	1							
OLE2	2.95	.63	.23	1						
OLE3	3.22	.46	.49	.36	1					
Con1	3.14	.48	.45	.46	.61	1				
Con2	3.17	.49	.42	.36	.56	.78	1			
Con3	3.15	.49	.41	.39	.54	.65	.76	1		
MS1	3.15	.49	.41	.32	.57	.46	.52	.48	1	
MS2	3.12	.53	.34	.22	.39	.42	.43	.44	.63	1

Note: All significant, $p < .01$.

The maximum likelihood estimation method was used for validating the measurement model and the structural model (Arbuckle 2011). The fit of the models was assessed using chi-square (χ^2) statistics and various fit indices such as the goodness of fit index (GFI), comparative fit index (CFI), the non-normed fit index (NFI), root mean square error of approximation (RMSEA) and standardized root mean square residual (SRMR). The GFI, CFI and NFI were all higher than .90, RMSEA was less than .08 and SRMR was less than .05. The indices have shown a good fit of the data to both the measurement model and structural model (Table 1.3).

Standardized factor loadings, composite reliability (CR) and average variance extracted (AVE) (Claes and Larcker 1981) were used to test the reliability and validity of the measurement model. The standardized factor loadings for all the factors were adequate and significant. The CR values have revealed that the factors were reliable and the AVE values have verified the validity of the factors. Therefore, the items were valid and reliable (Table 1.4).

1.4.3 Findings

The results have revealed that the online learning environment, concentration and metacognitive strategies significantly and positively correlated with one another. Online learning environment was positively and significantly associated with concentration and metacognitive strategies; concentration was positively and significantly associated with metacognitive strategies.

In answering the second research question, this means that the three factors were highly associated with one another and that all the hypotheses have been supported. This has led to findings that (1) a better online learning environment would allow the learners to better concentrate; (2) learners with good concentration would also make good use of metacognitive strategies during interpreter training; and (3) in other words, creating a good online learning environment, staying focused and concentrate and utilizing metacognitive strategies effectively would lead to more successful and effective interpreter training.

TABLE 1.3 Model fit indices

	N	χ^2	df	GFI	CFI	NFI	RMSEA	SRMR
				(>.90)	*(>.90)*	*(>.90)*	*(<.08)*	*(<.05)*
Measurement model	291	56.13***	17	.96	.98	.97	.079	.034
Structural model	291	56.13***	17	.96	.98	.97	.079	.034

***p < .001.

TABLE 1.4 Standardized factor loading for the measurement model

Items	Factor loading	Cronbach's alpha	CR	AVE
Online learning environment (OLE)		.60	.66	.40
1. During distance learning for interpreter training, if there is a sudden problem with the software or the internet, I will try to solve the problem or seek help.	.59			
2. In the interpreting class, when the teacher conducts distance teaching, I will shut down other web pages or programs on the computer (tablet/cellphone), so as not to interfere with my study.	.48			
3. Before participating in distance learning for the interpreting class, I will find an environment more suitable for learning (e.g., a place with stable internet speed and quiet surroundings).	.79			
Concentration (CON)		.89	.90	.74
1. During synchronous distance learning (i.e., the teacher and the students are online at the same time) in the interpreting class, I will still try to stay focused, despite unexpected problems with the equipment, software or the internet.	.85			
2. During synchronous distance learning (i.e., the teacher and the students are online at the same time) in the interpreting class, even if the teacher does not do a roll-call or does not ask us to turn on the camera to answer questions, I will still try to stay focused.	.91			
3. During synchronous distance learning (i.e., the teacher and the students are online at the same time) in the interpreting class, when other students are called upon to do, say, interpreting, I will still try to stay focused.	.82			

(Continued)

TABLE 1.4 (Continued)

Items	Factor loading	Cronbach's alpha	CR	AVE
Metacognitive strategies (MS)		.78	.79	.65
1. During asynchronous distance learning in the interpreting class (i.e., teachers pre-recorded videos), I tend to pause and re-watch passages that are not very clear.	.90			
2. During asynchronous distance learning in the interpreting class (i.e., teachers pre-recorded videos), I tend to pause at important passages and continue watching only after taking down notes or only after having gained a full understanding.	.70			

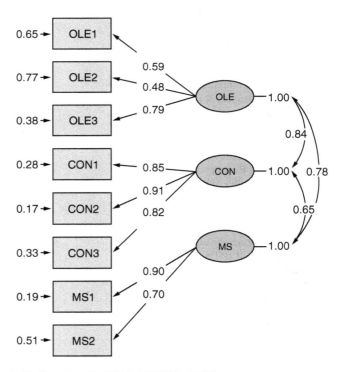

Chi–Square=47.55, df=17, P–value=0.00010, RMSEA=0.079

FIGURE 1.1   The measurement model and the structural model.

1.5   Discussion

1.5.1   *Effects on the three learning factors*

The study has revealed significant effects of online learning on concentration, learning environment and learning strategies. The results showed that online learning had a positive effect on the online learning environment, concentration and metacognitive learning strategies.

1.5.1.1   *Effect on online learning environment*

It is crucial for tertiary trainee interpreters to set up an optimal learning environment when learning online, due to the synchronous and asynchronous learning that takes place during interpreter training (Kiraly 2014). A suitable online learning environment is important as it can help reduce distractions, promote focus and concentration and allow for uninterrupted learning (González-Davies and Enríquez-Raído 2016).

An ideal online learning environment for tertiary trainee interpreters would have a stable internet connection, be free from distractions such as social media and be quiet so that learning can take place uninterrupted (Desjardins 2011). In order to promote effective learning and teaching during interpreter training, it is important to create a learning environment that is optimal for learning (Kiraly 2014).

1.5.1.2   *Effect on concentration*

Tertiary trainee interpreters need to stay laser-focused and concentrate even when technical glitches happen (González-Davies and Enríquez-Raído 2016) or even when other students are called upon to do interpreting (Kiraly and Hofmann 2016) in order to accomplish effective tertiary interpreter training using online learning during the pandemic. Concentration is an important skill for interpreters, as they need to be able to focus on the message that is being communicated, rather than getting distracted by other noise or activity happening around them (Kiraly and Hofmann 2016).

Additionally, it is worth noting that our findings have revealed that regardless of synchronous or asynchronous learning, students still remain concentration-intensive, as it, to a certain extent, requires students to self-regulate their own learning. Therefore, it is important for tertiary trainee interpreters to be aware of the importance of focus and how to manage their own concentration levels when working in online learning environments in association with interpreter training (Kiraly and Hofmann 2016; Robinson et al. 2016).

1.5.1.3   *Effect on metacognitive learning strategies*

As the pandemic continues, many tertiary institutions have had to rely increasingly on online learning (e.g., Ali 2020; Pokhrel and Chhetri 2021). This has presented a challenge for tertiary trainee interpreters, who need to be able to effectively manage their learning in order to make the most of these learning modalities (Pokhrel and Chhetri 2021).

Developing good metacognitive learning strategies is especially important for tertiary trainee interpreters because they often have to review teaching materials and take notes in a short period of time (Moser-Mercer 2015). By being aware of their own learning process, they can more effectively identify the areas where they need to focus their attention and make sure that they are retaining the information that they are learning (Moser-Mercer 2015).

While metacognitive learning strategies are important for all learners, they are especially crucial for tertiary trainee interpreters who are learning in an online mode. This is because these online learning modalities can be more challenging to navigate without a clear sense of how to effectively manage one's own learning (Motta 2016). That being said, by taking the time to develop good metacognitive learning strategies, tertiary trainee interpreters can ensure that they are getting the most out of their online learning experiences (Mahaye 2020).

1.5.2 Interplay between online learning environment, concentration and metacognitive strategies

The current study also sought to investigate the potential interrelationships between such three factors as the online learning environment, concentration and metacognitive strategies in tertiary interpreter training during the COVID-19 pandemic. The results have shown that there was a significant positive correlation between online learning environment, concentration and metacognitive strategies. The online learning environment was positively and significantly correlated with attention and metacognitive strategies. Attention was positively and significantly correlated with metacognitive strategies. All hypotheses have been supported and verified. This means that a better online learning environment allows learners to concentrate better. Additionally, highly focused learners also use metacognitive strategies well during interpreting training. The study has also suggested that creating a good online learning environment, maintaining focus and concentration and effectively using metacognitive strategies during the pandemic to a certain extent led to relatively more successful and effective tertiary interpreting training.

1.6 Limitations and conclusion

1.6.1 Limitations

The present study has several limitations. First, the sample size is moderately small and may not be representative of the population of tertiary trainee interpreters in the greater Chinese-speaking region. Second, the online questionnaire used in this study did not include a control group, which means that it is difficult to determine whether the findings can be generalized to other populations of trainee interpreters. Third, the online learning environment, concentration level and metacognitive learning strategies are complex educational psychology constructs that cannot be fully captured by the items in the questionnaire used in the present study. It is recommended that future studies use more comprehensive measures to assess these factors. Finally, the present study did not examine the long-term effects of online learning on trainee interpreters. It is further recommended that future research investigate the impact of online learning on trainee interpreters over a longer period of time.

1.6.2 Conclusion

The current study has important implications for both practice and research. In terms of practice, the findings suggest that educators need to create a good online learning environment, which can, in turn, help the learners sustain focus and concentration and utilize metacognitive strategies effectively. In terms of research, future studies could explore other factors that might impact tertiary interpreter training. Given the importance of online learning environment, concentration and metacognitive strategies in tertiary interpreter training, more research is needed to further understand their roles and effects.

To conclude, this study has exposed some key challenges that need to be addressed for online teaching to work effectively in the context of tertiary interpreter training. In particular, students need to be made aware of the importance of maintaining their concentration during online learning sessions and of using appropriate strategies to stay focused. Additionally, courses should be designed in a way that allows for sufficient interaction between students and instructors, as well as among the students themselves. If these concerns are addressed, then there is no doubt that online teaching can provide many benefits for T&I training programs, not only during exceptional circumstances like the COVID-19 pandemic but also as a helpful alternative to conventional face-to-face classroom teaching.

Note

1 In the present study, "tertiary T&I training" refers to a one-year translation or interpreting training course often provided as an elective module for undergraduate language majors. The study is primarily interested in those studying interpreting in this context.

References

Adedoyin, Olasile Babatunde, and Emrah Soykan. 2020. "Covid-19 Pandemic and Online Learning: The Challenges and Opportunities." *Interactive Learning Environments* 32: 1–13.

Al-Baadani, Ahmed Abdulkarem, and Mohammed Abbas. 2020. "The Impact of Coronavirus (Covid19) Pandemic on Higher Education Institutions (HEIs) in Yemen: Challenges and Recommendations for the Future." *European Journal of Education Studies* 7(7): 68–81.

Ali, Wahab. 2020. "Online and Remote Learning in Higher Education Institutes: A Necessity in Light of COVID-19 Pandemic." *Higher Education Studies* 10(3): 16–25.

Almahasees, Zakaryia, and Mutahar Qassem. 2021. "Faculty Perception of Teaching Translation Courses Online During Covid-19." *PSU Research Review* 6(3): 205–19.

Alwazna, Rafat Y. 2021. "Teaching Translation during COVID-19 Outbreak: Challenges and Discoveries." *Arab World English Journal* 12(4): 86–102.

Arbuckle, James L. 2011. *IBM SPSS Amos 20 User's Guide*. Amos Development Corporation, SPSS Inc.

Arkhipova, Maria V., Ekaterina E. Belova, Yulia A. Gavrikova, Natalya A. Lyulyaeva, and Elina D. Shapiro. 2018. "Blended Learning in Teaching EFL to Different Age Groups." In *International Conference on Humans,* edited by Elena G. Popkova, 380–86. Cham: Springer International Publishing.

Balakrishnan, Athira, Sandra Puthean, Gautam Satheesh, Unnikrishnan M. K., Muhammed Rashid, Sreedharan Nair, and Girish Thunga. 2021. "Effectiveness of Blended Learning in Pharmacy Education: A Systematic Review and Meta-analysis." *PLoS One* 16(6): e0252461. https://doi.org/10.1371/journal.pone.0252461.

Bergunde, Annika, and Sonja Pollabauer. 2019. "Curricular Design and Implementation of a Training Course for Interpreters in an Asylum Context." *Translation & Interpreting: The International Journal of Translation and Interpreting Research* 11(1): 1–21.

Biel, Łucja, and Vilelmini Sosoni. 2017. "The Translation of Economics and the Economics of Translation." *Perspectives* 25(3): 351–61.

Broadbent, Jaclyn. 2017. "Comparing Online and Blended Learner's Self-Regulated Learning Strategies and Academic Performance." *The Internet and Higher Education* 33: 24–32.

Brown, John Seely, Allan Collins, and Paul Duguid. 1989. "Situated Cognition and the Culture of Learning." *Educational Researcher* 18(1): 32–42. https://doi.org/10.3102/00346543018001032.

Bruffee, Kenneth A. 1999. *Collaborative Learning: Higher Education, Interdependence, and the Authority of Knowledge*. Baltimore: John Hopkins University Press.

Byrnes, Kevin G., Patrick A. Kiely, Colum P. Dunne, Kieran W. McDermott, and John Calvin Coffey. 2021. "Communication, Collaboration and Contagion: 'Virtualisation' of Anatomy during COVID-19." *Clinical Anatomy* 34(1): 82–89.

Cai, Rendong, and Yanping Dong. 2015. "Interpreter Training and Students of Interpreting in China." *Journal of Translation Studies* 16(4): 167–91.

Chang, Vincent Chieh-Ying. 2011. "Translation Directionality and Revised Hierarchical Model: An Eye-tracking Study." In *Cognitive Explorations of Translation*, edited by Sharon O'Brien, 154–74. London: Continuum.

Chang, Vincent Chieh-Ying 2024. Measuring Interpreting Learners' Cognitive Skills: Scale Validation Using Structural Equation Modeling. *Compilation & Translation Review* 17(2): 99–156.

Chang, Vincent Chieh-Ying, and I-Fei Chen. 2023. "Translation Directionality and the Inhibitory Control Model: A Machine Learning Approach to an Eye-tracking Study." *Frontiers in Psychology* 14: 1–12.

Chang, Vincent Chieh-Ying, Fabiana Gordon, Mark Shuttleworth, and Gabriela Saldanha. 2008. "Translators' Ocular Measures and Cognitive Loads during Translation." *Journal of Vision* 8(6): 647–47.

Chesterman, Andrew. 2009. "The Name and Nature of Translator Studies." *HERMES-Journal of Language and Communication in Business* 42: 13–22.

Class, Barbara, Barbara Moser-Mercer, and Kilian Seeber. 2004. "Blended Learning for Training Interpreter Trainers." In *3rd European Conference on E-learning*, edited by Dan Remenyi, 507–15. Reading: Academic Conference Ltd.

Crawford, Joseph, Kerryn Butler-Henderson, Jürgen Rudolph, Bashar Malkawi, Matt Glowatz, Rob Burton, Paola A. Magni, and Sophia Lam. 2020. "COVID-19: 20 Countries' Higher Education Intra-Period Digital Pedagogy Responses." *Journal of Applied Learning & Teaching* 3(1): 1–20.

Desjardins, Renée. 2011. "Facebook Me!: Initial Insights in Favour of Using Social Networking as a Tool for Translator Training." *Linguistica Antverpiensia, New Series—Themes in Translation Studies* 10: 175–193.

Dewey, John. 1986. "Experience and Education." *The Educational Forum* 50(3): 241–52.

Drane, Catherine, Lynette Vernon, and Sarah O'Shea. 2020. "The Impact of 'Learning at Home' on the Educational Outcomes of Vulnerable Children in Australia during the COVID-19 Pandemic." Literature Review Prepared by the National Centre for Student Equity in Higher Education. Curtin University, Australia.

Eser, Oktay, Miranda Lai, and Fatih Saltan. 2020. "The Affordances and Challenges of Wearable Technologies for Training Public Service Interpreters." *Interpreting* 22(2): 288–308.

Claes, Fornell, and David Larcker. 1981. "Evaluating Structural Equation Models with Unobservable Variables and Measurement Error." *Journal of Marketing Research, 18*(1), 39–50.

Galán-Mañas, Anabel, and Amparo Hurtado Albir. 2010. "Blended Learning in Translator Training: Methodology and Results of an Empirical Validation." *The Interpreter and Translator Trainer* 4(2): 197–231.

Gerbing, David W., and Anderson, James C. 1988. "An Updated Paradigm for Scale Development Incorporating Unidimensionality and Its Assessment." *Journal of Marketing Research* 25(2): 186–92.

Glasersfeld, Ernst von. 1988. "The Reluctance to Change a Way of Thinking." *The Irish Journal of Psychology* 9(1): 83–90.

González-Davies, Maria, and Vanessa Enríquez-Raído. 2016. "Situated Learning in Translator and Interpreter Training: Bridging Research and Good Practice." *The Interpreter and Translator Trainer* 10(1): 1–11.

Graham, Charles R. 2013. "Emerging Practice and Research in Blended Learning." *Handbook of Distance Education* 3: 333–50.

Graham, Charles R., Curtis R. Henrie, and Andrew S. Gibbons. 2013. "Developing Models and Theory for Blended Learning Research." *Blended Learning: Research Perspectives* 2: 13–33.

Guangul, Fiseha M., Adeel H. Suhail, Muhammad I. Khalit, and Basim A. Khidhir. 2020. "Challenges of Remote Assessment in Higher Education in the Context of COVID-19: A Case Study of Middle East College." *Educational Assessment, Evaluation and Accountability* 32(4): 519–35.

Haider, Ahmad S., and Saleh Al-Salman. 2020. "COVID-19'S Impact on the Higher Education System in Jordan: Advantages, Challenges, and Suggestions." *Humanities & Social Sciences Reviews* 8(4): 1418–28.

Halverson, Lisa R., and Charles R. Graham. 2019. "Learner Engagement in Blended Learning Environments: A Conceptual Framework." *Online Learning* 23(2): 145–78.

Hansen, Inge Gorm, and Miriam Shlesinger. 2007. "The Silver Lining: Technology and Self-study in the Interpreting Classroom." *Interpreting* 9(1): 95–118.

Holmes, James S. 2021. *Translated!: Papers on Literary Translation and Translation Studies. With an introduction by Raymond van den Broeck*. Amsterdam: Brill.

Ismail, Rosmawati, Zaleha Ismail, and Yudariah Mohammad Yusof. 2016. "Blended Learning Environment in Tertiary Education: A Meta-Analysis." *Advanced Science Letters* 22(12): 4263–66.

Jiang, Haozhe, AYM Islam, Xiaoqing Gu, and Jonathan Michael Spector. 2021. "Online Learning Satisfaction in Higher Education during the COVID-19 Pandemic: A Regional Comparison Between Eastern and Western Chinese Universities." *Education and Information Technologies* 26(6): 6747–69.

Kelly, Dorothy, and Anne Martin. 2019. "Training and Education, Curriculum." In *Routledge Encyclopedia of Translation Studies*, edited by Mona Baker and Gabriela Saldanha, 591–96. Abingdon: Routledge.

Kelvin, David J., and Salvatore Rubino. 2020. "Fear of the Novel Coronavirus." *The Journal of Infection in Developing Countries* 14(1): 1–2.

Kiraly, Don, and Sascha Hofmann. 2016. "Towards a Post-Positivist Curriculum Development Model for Translator Education." In *Towards Authentic Experiential Learning in Translator Education*, edited by Don Kiraly, Silvia Hansen-Schirra, and Karin Maksymski, 67–88. Göttingen: V&R Press.

Kiraly, Donald. 2014. *A Social Constructivist Approach to Translator Education: Empowerment from Theory to Practice*. London: Routledge.

Kiraly, Donald. 2015. "Occasioning Translator Competence: Moving Beyond Social Constructivism toward a Postmodern Alternative to Instructionism." *Translation and Interpreting Studies. The Journal of the American Translation and Interpreting Studies Association* 10(1): 8–32.

Ko, Leong. 2006. "Teaching Interpreting by Distance Mode: Possibilities and Constraints." *Interpreting* 8(1): 67–96.

Kramsch, Claire. 2014. "Language and Culture." *AILA Review* 27(1): 30–55. https://doi.org/10.1075/aila.27.02kra.

Kumar, Yukteshwar. 2017. "Chinese Interpreting Programmes and Pedagogy." In *The Routledge Handbook of Chinese Translation*, edited by Chong Shei and Zhao-Ming Gao, 307–20. New York: Routledge.

Lane, Stephen, John G. Hoang, Jacqueline P. Leighton, and Anna Rissanen. 2021. "Engagement and Satisfaction: Mixed-Method Analysis of Blended Learning in the Sciences." *Canadian Journal of Science, Mathematics and Technology Education* 21(1): 100–22.

Laviosa, Sara. 2014. *Translation and Language Education: Pedagogic Approaches Explored*. London: Routledge.

Lee, Ting Ying, and Po Sen Liao. 2010. "Assessing College Student Learning in Interpretation Courses." *Studies of Translation and Interpretation* 13: 255–92.

Leonardi, Vanessa. 2010. *The Role of Pedagogical Translation in Second Language Acquisition: From Theory to Practice*. Bern: Peter Lang.

Lin, Oscar Chun-hung. 2015. "The Translation Industry in Taiwan in the Context of Globalisation: Facing the Development of Professional Translation and Master of Translation and Interpreting." In *Translation and Cross-Cultural Communication Studies in the Asia Pacific*, edited by Leong Ko, 369–87. Berlin: Brill.

Liu, Jie. 2020. *Interpreter Training in Context*. Singapore: Springer.

Liu, Kanglong, Hao Yin, and Andrew KF Cheung. 2024. "Interactional Metadiscourse in Translated and Non-translated Medical Research Article Abstracts: A Corpus-assisted Study." *Perspectives* 33: 1–21.

Mahaye, Ngogi Emmanuel. 2020. "The Impact of COVID-19 Pandemic on Education: Navigating Forward the Pedagogy of Blended Learning." *Research Online* 5(1): 4–9.

Malmkjær, Kirsten. 2010. "Language Learning and Translation." In *Handbook of Translation Studies*, edited by Yves Gambier and Luc van Doorslaer, 185–90. Amsterdam: John Benjamins.

Marita, Yosi, and Elva Utami. 2020. "The Implementation of Blended Learning in English Learning." *International Conference on the Teaching English and Literature* 1(1): 257–63.

Martins, Paulo, Henrique Rodrigues, Tânia Rocha, Manuela Francisco, and Leonel Morgado. 2015. "Accessible Options for Deaf People in E-learning Platforms: Technology Solutions for Sign Language Translation." *Procedia Computer Science* 67: 263–72.

Massey, Gary. 2005. "Process-Oriented Translator Training and the Challenge for E-learning." *Meta: Journal des Traducteurs/Meta: Translators' Journal* 50(2): 626–33.

McInerney, Dennis M. 2013. *Educational Psychology: Constructing Learning*. Parkside, SA: Pearson Higher Education AU.

Melchor, María Dolores Rodríguez. 2018. "Pedagogical Assistance for the XXI Century: The Interaction Between DG-SCIC, DG-INTE and Universities in the Field of Blended Learning for Interpreter Training." *CLINA: An Interdisciplinary Journal of Translation, Interpreting and Intercultural Communication* 4(1): 89–103.

Mishra, Lokanath, Tushar Gupta, and Abha Shree. 2020. "Online Teaching-Learning in Higher Education during Lockdown Period of COVID-19 Pandemic." *International Journal of Educational Research Open* 1: 100012.

Mok, Ka Ho, Weiyan Xiong, and Huiyuan Ye. 2021. "COVID-19 Crisis and Challenges for Graduate Employment in Taiwan, Mainland China and East Asia: A Critical Review of Skills Preparing Students for Uncertain Futures." *Journal of Education and Work* 34(3): 247–61.

Moorhouse, Benjamin Luke. 2020. "Adaptations to a Face-to-Face Initial Teacher Education Course 'Forced' Online Due to the COVID-19 Pandemic." *Journal of Education for Teaching* 46(4): 609–11.

Moser-Mercer, Barbara. 2008. "Skill Acquisition in Interpreting: A Human Performance Perspective." *The Interpreter and Translator Trainer* 2(1): 1–28.

Moser-Mercer, Barbara. 2015. *The Routledge Handbook of Interpreting*. London: Routledge.

Moser-Mercer, Barbara, Barbara Class, and Kilian Seeber. 2005. "Leveraging Virtual Learning Environments for Training Interpreter Trainers." *Meta: Journal des Traducteurs/Meta: Translators' Journal* 50(4). https://doi.org/10.7202/019872ar.

Motta, Manuela. 2016. "A Blended Learning Environment Based on the Principles of Deliberate Practice for the Acquisition of Interpreting Skills." *The Interpreter and Translator Trainer* 10(1): 133–49.

Mouzourakis, Panayotis. 2006. "Remote Interpreting: A Technical Perspective on Recent Experiments." *Interpreting* 8(1): 45–66.

Mozelius, Peter. 2020. "Post Corona Adapted Blended Learning in Higher Education." In *Responding to Covid-19: The University of the Future*, edited by Dan Remenyi, Ken A Grant, and Shawren Singh, 1. Reading: ACIL.

Nkengasong, John. 2020. "China's Response to a Novel Coronavirus Stands in Stark Contrast to the 2002 SARS Outbreak Response." *Nature Medicine* 26(3): 31011.

Pan, Jun, and Jackie Xiu Yan. 2012. "Learner Variables and Problems Perceived by Students: An Investigation of a College Interpreting Programme in China." *Perspectives* 20(2): 199–218.

Perez, E., and S. Hodakova. 2021. "Translator and Interpreter Training During the COVID-19 Pandemic: Procedural, Technical and Psychosocial Factors in Remote Training." *Current Trends in Translation Teaching and Learning E* 8: 276–312. https://doi.org/10.51287/cttle20219.

Pintrich, Paul R. 1991. "Motivated Strategies for Learning Questionnaire (MSLQ) Manual." *National Center for Research to Improve Postsecondary Teaching and Learning*, 1–79.

Pokhrel, Sumitra, and Roshan Chhetri. 2021. "A Literature Review on Impact of COVID-19 Pandemic on Teaching and Learning." *Higher Education for the Future* 8(1): 133–41.

Pokorn, Nike K. 2009. "Natives or Non-Natives? That Is the Question… Teachers of Translation into Language B." *The Interpreter and Translator Trainer* 3(2): 189–208.

Presas, Marisa. 2012. "Training Translators in the European Higher Education Area: A Model for Evaluating Learning Outcomes." *The Interpreter and Translator Trainer* 6(2): 139–69.

Pym, Anthony, and Nune Ayvazyan. 2017. "Linguistics, Translation and Interpreting in Foreign-Language Teaching Contexts." In *The Routledge Handbook of Translation Studies and Linguistics*, edited by Kirsten Malmkjær, 393–407. New York: Routledge.

Pym, Anthony, Kirsten Malmkjær, and M. Gutiérrez-Colón. 2013. "Translation and Language Learning: The Role of Translation in the Teaching of Languages in the European Union." *Publications Office of the European Union* 15: 1–102.

Resien, Resien, Harun Sitompul, and Julaga Situmorang. 2020. "The Effect of Blended Learning Strategy and Creative Thinking of Students on the Results of Learning Information and Communication Technology by Controlling Prior Knowledge." *Budapest International Research and Critics in Linguistics and Education* 3(2): 879–93.

Risku, Hanna. 2016. "Situated Learning in Translation Research Training: Academic Research as a Reflection of Practice." *The Interpreter and Translator Trainer* 10(1): 12–28.

Robinson, Bryan J., María Dolores Olvera-Lobo, and Juncal Gutiérrez-Artacho. 2016. "After Bologna: Learner-and Competence-Centred Translator Training for 'Digital Natives'." *From the Lab to the Classroom and Back Again: Perspectives on Translation and Interpreting Training*, edited by Celia Martín de León, González Ruiz and Víctor Manuel, 2: 325–59. Lausanne: Peter Lang.

Rorty, Richard. 2009. *Philosophy and the Mirror of Nature*. Princeton University Press.

Şahin, Mehmet. 2013. "Virtual Worlds in Interpreter Training." *The Interpreter and Translator Trainer* 7(1): 91–106.

Sandrelli, Annalisa. 2002. "Computers in the Training of Interpreters: Curriculum Design Issues." *Computers in the Training of Interpreters: Curriculum Design Issues* 12: 1000–16.

Sandrelli, Annalisa, and Jesus De Manuel Jerez. 2007. "The Impact of Information and Communication Technology on Interpreter Training: State-of-the-art and Future Prospects." *The Interpreter and Translator Trainer* 1(2): 269–303.

Setton, Robin. 2011. *Interpreting Chinese, Interpreting China*. Amsterdam: John Benjamins.

Shim, Tae Eun, and Song Yi Lee. 2020. "College Students' Experience of Emergency Remote Teaching Due to COVID-19." *Children and Youth Services Review* 119: 105578.

Singh, Harvey. 2021. "Building Effective Blended Learning Programs." In *Challenges and Opportunities for the Global Implementation of E-Learning Frameworks*, edited by Badrul H. Khan, Saida Affouneh, and Soheil Hussein Salha, 15–23. Hershey, PA: IGI Global.

Sitthiworachart, Jirarat, Mike Joy, and Jon Mason. 2021. "Blended Learning Activities in an e-Business Course." *Education Sciences* 11(12): 1–16.

Sohrabi, Catrin, Ginimol Mathew, Thomas Franchi, Ahmed Kerwan, Michelle Griffin, Jennick Soleil C. Del Mundo, Syed Ahsan Ali, Maliha Agha, and Riaz Agha. 2021. "Impact of the Coronavirus (COVID-19) Pandemic on Scientific Research and Implications for Clinical Academic Training–A Review." *International Journal of Surgery* 86: 57–63.

Tao, Youlan. 2016. "Translator Training and Education in China: Past, Present and Prospects." *The Interpreter and Translator Trainer* 10(2): 204–23.

Thomas, Aliki, Anita Menon, Jill Boruff, Ana Maria Rodriguez, and Sara Ahmed. 2014. "Applications of Social Constructivist Learning Theories in Knowledge Translation for Healthcare Professionals: A Scoping Review." *Implementation Science* 9(1): 1–20.

Toquero, Cathy Mae. 2020. "Challenges and Opportunities for Higher Education Amid the COVID-19 Pandemic: The Philippine Context." *Pedagogical Research* 5(4): em0063. https://doi.org/10.29333/pr/7947.

Toury, Gideon. 1995. *Descriptive Translation Studies and Beyond*. Amsterdam: John Benjamins.

Van Doorslaer, Luc. 2007. "Risking Conceptual Maps: Mapping as a Keywords-related Tool Underlying the Online Translation Studies Bibliography." *Target: International Journal of Translation Studies* 19(2): 217–33.

Varney, Jennifer. 2009. "From Hermeneutics to the Translation Classroom: A Social Constructivist Approach to Effective Learning." *Translation & Interpreting: The International Journal of Translation and Interpreting Research* 1(1): 29–45.

Viner, Russell M., Simon J. Russell, Helen Croker, Jessica Packer, Joseph Ward, and Claire Stansfield. 2020. "School Closure and Management Practices during Coronavirus Outbreaks Including COVID-19: A Rapid Systematic Review." *The Lancet Child & Adolescent Health* 4(5): 397–404.

Vygotsky, Lev S. 1994. "Extracts from Thought and Language and Mind in Society." In *Language, Literacy and Learning in Educational Practice*, edited by Barry Stierer, and Janet Maybin, 45–58. Clevedon: Multilingual Matters.

Wang, Binhua, and Mu Lei. 2009. "Interpreter Training and Research in Mainland China: Recent Developments." *Interpreting* 11(2): 267–83.

Williams, Jenny, and Andrew Chesterman. 2002. *The Map: A Beginner's Guide to Doing Research in Translation Studies*. London: St. Jerome Publishing.

Wuwung, Olivia Cherly, and Jeane Marie Tulung. 2021. "How To Develop A Christian Religious Education Learning Model With A Blended Learning Approach (Need Analysis Studies)." *International Journal of Education, Information Technology, and Others* 4(1): 1–5.

Xiao, Chunchen, and Yi Li. 2020. "Analysis on the Influence of the Epidemic on the Education in China." In *2020 International Conference on Big Data and Informatization Education (ICBDIE)*, 143–47. Zhangjiajie, China.

Xu, Jianzhong. 2005. "Training Translators in China." *Meta: Journal des Traducteurs/Meta: Translators' Journal* 50(1): 231–49.

Yan, Jackie Xiu, Jun Pan, and Honghua Wang. 2017. *Research on Translator and Interpreter Training: A Collective Volume of Bibliometric Reviews and Empirical Studies on Learners*. Cham: Springer.

Yang, Xiaoming, Xing Zhou, and Jie Hu. 2021. "Students' Preferences for Seating Arrangements and Their Engagement in Cooperative Learning Activities in College English Blended Learning Classrooms in Higher Education." *Higher Education Research & Development* 41(2): 1–16.

Zanettin, Federico, Gabriela Saldanha, and Sue-Ann Harding. 2015. "Sketching Landscapes in Translation Studies: A Bibliographic Study." *Perspectives* 23(2): 161–82.

Zhan, Cheng. 2014. "Professional Interpreter Training in Mainland China: Evolution and Current Trends." *International Journal of Interpreter Education* 6(1): 2.

Zhong, Yong. 2017. "Global Chinese Translation Programmes: An Overview of Chinese English Translation/Interpreting Programmes." In *The Routledge Handbook of Chinese Translation*, edited by Chong Shei and Zhao-Ming Gao, 19–36. New York: Routledge.

APPENDIX 1

Questionnaire

1 Gender: Female Male
2 Age:
3 How many years have you spent learning English as a second language?
4 How many hours of interpreting training do you do in your university per week?
5 In which city is your university located?
6 During distance learning for interpreter training, if there is a sudden problem with the software or the internet, I will try to solve the problem or seek help.
7 In the interpreting class, when the teacher conducts distance teaching, I will shut down other web pages or programs on the computer (tablet/cellphone), so as not to interfere with my study.
8 Before participating in distance learning for the interpreting class, I will find an environment more suitable for learning (e.g., a place with stable internet speed and quiet surroundings).
9 During synchronous distance learning (i.e., the teacher and the students are online at the same time) in the interpreting class, I will still try to stay focused, despite unexpected problems with the equipment, software or the internet.
10 During synchronous distance learning (i.e., the teacher and the students are online at the same time) in the interpreting class, even if the teacher does not do a roll-call or does not ask us to turn on the camera to answer questions, I will still try to stay focused.
11 During synchronous distance learning (i.e., the teacher and the students are online at the same time) in the interpreting class, when other students are called upon to do, say, interpreting, I will still try to stay focused.
12 During asynchronous distance learning in the interpreting class (i.e., teachers pre-recorded videos), I tend to pause and re-watch passages that are not very clear.
13 During asynchronous distance learning in the interpreting class (i.e., teachers pre-recorded videos), I tend to pause at important passages and continue watching only after taking down notes or only after having gained a full understanding.

2

A POST-COVID REFLECTION OF DISTANCE INTERPRETING IN TAIWAN

Damien Chiaming Fan

2.1 Introduction

The field of interpreting has undergone significant transformations in recent years, particularly in response to technological advancements and global events. The advent of distance interpreting (DI) has revolutionized the way interpreters work, presenting new opportunities and challenges. This shift has been further accelerated by the COVID-19 pandemic, which necessitated widespread adoption of DI practices. As the profession adapts to these changes, it is crucial to examine how interpreters' experiences and perceptions of DI have evolved over time. This study aims to contribute to this understanding by investigating the situation in Taiwan, building upon previous research and addressing gaps in longitudinal studies on this topic.

Several studies examined interpreters' performances, experiences and perceptions of DI prior to the COVID-19 pandemic. Braun (2017), Moser-Mercer (2003), Mouzourakis (2006), Roziner and Shlesinger (2010) and Seeber et al. (2019) conducted research using varied methodologies to investigate this topic. While Braun (2017) employed micro-analysis of recorded interpretations and Seeber et al. (2019) combined online questionnaires with structured interviews, the other studies utilized experimental designs comparing remote and on-site interpreting conditions. A common finding across studies was that interpreters reported feeling more fatigued, stressed and alienated when interpreting remotely compared to on-site (Moser-Mercer 2003; Mouzourakis 2006; Roziner and Shlesinger 2010). However, the studies differed in their findings regarding interpreting quality. Moser-Mercer (2003) found a faster decline in quality for DI, while Roziner and Shlesinger (2010) observed only a slight, non-significant decrease. Interestingly, Roziner and Shlesinger (2010) and Seeber et al. (2019) noted a discrepancy between interpreters' subjective perceptions of poorer performance remotely and the objective quality

DOI: 10.4324/9781003597711-3

assessments showing minimal differences. The studies also highlighted technological issues, particularly related to sound quality and synchronization of audio and video, as key challenges in DI (Braun 2017; Mouzourakis 2006; Seeber et al. 2019). Overall, these pre-pandemic studies revealed interpreters' general preference for on-site over DI, while identifying various ergonomic, cognitive and technological factors influencing the DI experience. However, Seeber et al. (2019) found that with high-quality technical setups and support, interpreters' attitudes toward DI could be more positive than previously reported.

The onset of the COVID-19 pandemic necessitated a rapid shift toward DI, prompting further research on the topic. In addition to entire volumes dedicated to a variety of topics such as training (Cheung 2022a, 2022b; Liu and Cheung 2023), several studies conducted during this period focused on specific markets, including Taiwan (Fan 2022), Turkey (Kincal and Ekici 2020), Spain (Mahyub Rayaa and Martin 2022) and Japan (Matsushita 2022). Buján and Collard (2022) conducted a larger study, involving 946 respondents from 7 regions and 19 countries. These studies typically employed mixed-method designs that utilized survey questionnaires and interviews. Their results corroborated earlier findings regarding challenges with DI, such as technical issues, increased cognitive load and feelings of isolation. However, they also noted some benefits, including reduced travel and improved work-life balance (Buján and Collard 2022; Fan 2022; Matsushita 2020). A consistent finding across both pre- and post-pandemic studies was interpreters' general preference for on-site over remote interpreting. The pandemic-era research also revealed a growing acceptance of remote interpreting as an inevitable part of the profession's future, despite ongoing reservations (Buján and Collard 2022; Fan 2022; Mahyub Rayaa and Martin 2022).

In addition to questionnaires and interviews, recent studies have employed other methodologies to examine interpreting performance in the DI setting. Cheung (2022a, 2022b), Chmiel and Spinolo (2022) and Hale et al. (2022) utilized experimental designs, albeit with different focuses. Cheung conducted an experiment comparing DI performance at home and hub settings, specifically examining numerical accuracy in English-Chinese interpretation. The study found that while overall accuracy rates were similar between the two settings, interpreters in the hub setting adopted approximation strategies more extensively when dealing with complex numbers and could be more strategic, possibly due to a lower perceived cognitive load. Chmiel and Spinolo's approach combined eye-tracking, performance measures and questionnaires to assess interpreter experience and performance across various DI configurations. They suggested that the presence and type of boothmate (co-located, non-co-located using chat or virtual booth) and the multimodal input channels significantly impact interpreters' performance and experience in remote simultaneous interpreting (RSI). Hale et al. (2022) compared the performance of interpreters in face-to-face, video remote and audio remote settings during simulated police interviews, finding that face-to-face and video remote interpreting resulted in similar accuracy and effectiveness, while audio remote interpreting was less effective. Interpreters overwhelmingly preferred face-to-face interpreting due to better non-verbal communication and fewer technical issues.

In contrast, Donovan (2023) and Frittella and Rodríguez (2022) employed observational methods. Donovan analyzed interpreter interactions during remote and on-site assignments, revealing that remote settings significantly reduced certain forms of interpreter interaction and cooperation. This reduction potentially threatens professional cohesion and limits opportunities for knowledge sharing among interpreters. Frittella and Rodríguez (2022) conducted usability testing of a specific remote interpreting platform with professional interpreters, identifying key user interface features that influenced system usability and participant satisfaction. Their findings emphasized the importance of simplicity in interface design and the need for features that enable naturalistic interaction between booth partners.

Despite the methodological differences, these studies collectively highlight the complexities introduced by DI, including technological challenges and altered communication dynamics among interpreters, and the need to investigate the issue of cognitive load (Chmiel and Spinolo 2022; Zhu and Aryadoust 2022). They also point to potential benefits, such as integrated computer-assisted interpreting tools, while underscoring the importance of further research to optimize DI setups and develop targeted training for interpreters in this modality.

The broader implications of DI on the interpreting profession have also been analyzed and theorized by some studies. A common thread among them is the concern over the commodification and potential devaluation of interpreting services in the digital age. Cronin and Delgado Luchner (2021) framed DI within the context of cognitive capitalism, arguing it represents a form of exploitation of interpreters' communication labor that threatens professional cohesion and autonomy. Hoyte-West (2022) further explored these themes, focusing on the pandemic's impact on professional status, particularly in the context of the European Union, noting challenges faced by freelancers and accelerated technological changes. Building on these broader sociological perspectives, Özkaya Marangoz (2023) introduced the concept of "uberization" in interpreting, highlighting risks such as the erosion of quality standards and the commodification of specialized skills. She argued that while technology can enhance the profession, it is crucial to maintain professionalism, ethics and specialized skills fundamental to effective interpreting. This perspective aligns with Giustini's (2024) application of labor process theory, which examines how digital platforms are reshaping business models and working conditions for interpreters. Collectively, these studies reveal deeper undercurrents of and more sustained impacts on the interpreting profession as its stakeholders adapt to new technological realities. Whether interpreters, who are the central characters in this ongoing development, are cognizant of the invisible forces that are changing how they are identified, employed and perceived remains to be explored.

Despite this growing body of research, there remains a notable gap in the lack of longitudinal studies examining the evolution of interpreters' experiences and perceptions of DI over time. Most studies have provided snapshots of the situation at specific points during the pandemic, but few have traced how these perceptions and practices have changed as interpreters gained more experience with DI. To address this gap, the present study aims to answer the following research questions:

1 How have the experiences and perceptions of DI evolved among Taiwan's interpreters?
2 What are the major advantages, disadvantages and concerns they have about DI after gaining substantial experience in this mode?

To answer these questions, a mixed-method approach incorporating a survey questionnaire and semi-structured interviews was adopted. This study builds upon Fan's (2022) research, providing a comparative perspective on how interpreters' views and practices have developed over time. For clarity and consistency, it is essential to define the key modes of interpreting examined in this study: On-site interpreting refers to a setup where interpreters, colleagues, speakers and audience are physically co-located, with audio transmitted to interpreters via physical cables. DI, in contrast, involves interpreters, colleagues, speakers and audience in separate locations, with audio and video transmitted via the internet to interpreters, and interpretations delivered to the audience through the same medium. RSI, a dominant mode of interpretation delivery during the COVID-19 pandemic, is categorized under DI. Hybrid interpreting represents a combination where some participants are on-site while others are online, but crucially, all audio and video are transmitted via the internet to interpreters, and their interpretations are likewise internet delivered (Fan 2022).

By examining these different modes and their implications, this study hopes to contribute to a better understanding of the longer-term implications of DI for the interpreting profession and helps inform future practices and policies in this rapidly evolving field.

2.2 Methods

2.2.1 Questionnaire

The questionnaire in Chinese was developed using SurveyCake, a cloud-based survey tool that facilitates participation via both PC and mobile devices. It comprised 23 items organized into four main sections: background information, workload, working conditions and perception of DI.

Questions 1–3 gathered participants' years of experience, age group and working languages. Questions 4–6 focused on participants' workload patterns across three modes of delivery: on-site, distance and hybrid interpreting. The survey examined these modes over three distinct time periods: before the COVID-19 border closure, during the pandemic and after Taiwan reopened its borders. For the pre-pandemic period, participants were asked about their annual average workload before

February 2020. The pandemic period covered March 2020 to October 2022, during which border restrictions were implemented. The post-COVID period spanned November 2022 to March 2024. For these latter two periods, participants reported their total number of working days. To protect privacy and facilitate responses, the survey offered six predefined workload ranges instead of requesting exact figures. Participants selected the appropriate range for each mode and time period, with these ranges detailed in Table 2.1. This approach provided a structured yet discreet method for capturing the evolution of interpreters' work patterns throughout the COVID-19 pandemic. To facilitate the comparison of workload patterns across different delivery modes and time periods, the ordinal data were converted to a weighted average score. This conversion allows for a more intuitive interpretation of the relative prevalence of each delivery mode over time. The ordinal categories were treated as a continuous scale, with each category assigned a value from 1 to 6, corresponding to the lowest to highest workload ranges. The weighted average was then calculated by multiplying each scale value by the number of respondents in that category, summing these products and dividing by the total number of respondents. While this approach introduces some imprecision, particularly for open-ended categories, it provides a single, comparable value for each delivery mode and time period, enabling a clearer visualization of trends. This method is similar to that used in Likert scale conversions (Harpe 2015) and allows for a more accessible representation of the data while maintaining the original ordinal categories for reference.

Question 7 asked how often participants encountered changes in delivery mode (DI to on-site, DI to hybrid or continued DI) after border reopening. Question 8 inquired whether new clients emerged due to the advantages of DI.

Questions 9–15 focused on various aspects of working conditions, including whether participants quoted hourly, half-day or daily rates; whether quotes changed; the presence of additional or modified contractual terms; duration of DI work; team strength; location of booth mates; and the medium used for DI delivery.

TABLE 2.1 Workload options for each period

Pre-COVID *Annual average before February 2020*	*During COVID* *Total number of days of work from March 2020 to October 2022 (32 months)*	*Post-COVID* *Total number of days of work from November 2022 to March 2024 (17 months)*
Less than 20 days	Less than 40 days	Less than 30 days
21–40 days	41–80 days	31–60 days
41–60 days	81–120 days	61–90 days
61–80 days	121–160 days	91–120 days
81–100 days	161–200 days	121–150 days
More than 100 days	More than 200 days	More than 150 days

Questions 16–22 asked participants to select the three most significant advantages of DI from a list of eight options (or specify others) and to choose the five most significant drawbacks from a list of 13 options (or specify others). Participants were also asked whether they had ever declined DI assignments, their overall impression of DI (positive or negative) and their ranked preference for on-site, DI and hybrid modes. The final question invited willing participants to provide their email addresses for follow-up interviews.

2.2.2   Procedure

The study employed a combination of convenience and snowball sampling methods to maximize participation. A list of interpreters based in Taiwan was compiled, comprising members of the International Association of Conference Interpreters ($n = 8$), members of a closed interpreters' group on Facebook ($n = 74$) and members of a private interpreters' group chat on a social media messaging application ($n = 60$). Due to data privacy restrictions, it was difficult to discern overlapping memberships among these groups. To broaden the reach, the questionnaire link was also sent to interpreter trainers who taught in postgraduate interpreting programs in Taiwan and posted on two closed alumni Facebook group pages of postgraduate interpreting programs in Taiwan. Snowball sampling was employed by encouraging participants to share the survey link on various social media platforms. Eligibility criteria included being domiciled in Taiwan, having experience performing both on-site and RSI, working as a freelance spoken language interpreter and relying on interpreting as the main source of income. Additionally, participants must have started their interpreting career before the pandemic began, effectively excluding those who entered the profession in or after 2020. Between March 28 and April 15, 2024, a total of 40 responses were collected, which was one fewer than the number reported in Fan (2022).

In July 2024, the results of the questionnaire survey were compiled into an executive summary report and sent to 16 respondents who had expressed willingness to participate in follow-up interviews by providing their contact email addresses. Of these, 15 were available and subsequently contacted to schedule interviews in July 2024. Informed consent was obtained before conducting the interview sessions via video conferencing in Mandarin Chinese. Each session lasted between 40 and 70 minutes.

2.2.3   Interview content and analysis

The semi-structured interviews were designed to correspond with the three broad themes of the survey questionnaire: workload patterns, working conditions and perceptions of DI. The questions aimed to delve deeper into the rationale behind participants' survey responses. Regarding workload patterns, participants were asked to expound on the changes in the proportion of different delivery modes across the three time periods. They were also prompted to share their views on why clients or organizers opted for particular delivery modes. Concerning working conditions, participants were asked to explain why they proactively requested different contractual terms for DI or why they did not deem changes necessary. They were also invited to share any specific working habits they developed to cope with DI. In terms of DI perception, participants were asked to elaborate on the advantages and disadvantages of DI and to discuss any impacts on their mental and physical health. Finally, participants were asked to comment on whether and how DI changed the role and status of interpreters, as well as to share their thoughts on the future development of DI.

The content of the interview sessions was recorded, transcribed verbatim, translated to English by the researcher and subsequently cleaned for grammar and clarity with the assistance of Claude, an AI language model (Anthropic 2024). The transcripts were then subjected to thematic content analysis using the computer-aided qualitative data analysis software ATLAS.ti 24.1.1 for Mac (ATLAS.ti Scientific Software Development GmbH 2023). An abductive approach was used, combining both deductive and inductive methods to gain a comprehensive understanding of the data (Proudfoot 2022; Thompson 2022). This method has also been used to analyze the experiences of health workers during the COVID-19 pandemic (Hennein and Lowe 2020). Initially, deductive coding was applied using a set of pre-determined codes derived from existing literature on DI. These codes reflected key concepts, issues and findings from prior research in the field. Simultaneously, inductive coding was used to allow new themes to emerge organically from the data, capturing patterns, concepts and categories directly from the interviews that may not have been previously identified in the literature. This abductive approach enabled a continuous interplay between prior research findings and new data. When findings emerged that were not accounted for by existing research, new codes were developed. Conversely, when the data aligned with previous findings, the pre-existing codes were applied. This process allowed for both the application of established knowledge in the field and the discovery of novel insights specific to the post-COVID context. Throughout the analysis, constant comparison was used to identify similarities and differences across participants' responses, helping to ensure the robustness of the emerging themes. This combined approach enabled the integration of inductive insights with the deductive framework based on prior research, refining and redefining themes iteratively. Table 2.2 presents the final coding scheme employed for analyzing the interview content, reflecting this integration of literature-driven and data-driven codes.

2.3 Results

2.3.1 Participants and interviewees

A total of 40 participants completed the questionnaire. The age distribution shows that the largest group was between 36 and 45 years old ($n = 18$, 45.0%), followed

TABLE 2.2 Coding scheme

Theme	Category	Code
Workload patterns	Changes in interpreting modes over time	Pre-pandemic distribution
		During pandemic distribution
		Post-pandemic distribution
	Factors influencing client choices	Type and nature of meetings
		Industry-specific preferences
		Cost and other considerations
Working conditions	Technical aspects	Equipment and setup requirements
		Platform usage and preferences
		Audio quality issues
	Preparation and execution	Changes in preparation methods
		Use of multiple devices and screens
		Reliance on subtitles and other aids
	Interpersonal dynamics	Interaction with clients and audiences
		Collaboration with interpreting colleagues
		RSI platforms vs general video conferencing platforms
	Contractual considerations	Fees and rates
		Working hours
		New clauses related to DI
Perceptions of DI	Advantages	Time and cost savings
		Work-life balance
		Increased job opportunities
	Disadvantages	Technical challenges and stress
		Lack of personal interaction
		Potential health impacts
	Impact on interpreter roles and status	Expanded responsibilities
		Concerns about diminished presence
	Change of perception	Acclimatization
		Potential impact of AI

by those aged 46–55 ($n = 13$, 32.5%). Together, these two age groups account for 77.5% of the respondents.

Regarding years of experience, the most common range was 11–15 years ($n = 14$, 35.0%), followed by those with more than 20 years of experience ($n = 12$, 30.0%). In terms of working languages, the majority of respondents ($n = 26$, 65.0%) worked between Mandarin Chinese and English. The next most common working language in addition to Mandarin Chinese was Japanese or Korean ($n = 11$, 27.5%). A detailed breakdown of age, years of experience and working languages in addition to Mandarin Chinese can be found in Table 2.3.

Among the 15 interviewees, the vast majority worked with Mandarin Chinese and English, one worked with French and one worked with Japanese. The only two interviewees who identified as male were P2 and P6. Table 2.4 provides a detailed profile of each interviewee, including their age group, years of experience and their language pair.

TABLE 2.3 Demographic characteristics of survey respondents ($n = 40$)

Category	Group	Frequency (N)	Percentage (%)
Age	26–35	3	7.5
	36–45	18	45.0
	46–55	13	32.5
	56–65	6	15.0
Years of experience	6–10	8	20.0
	11–15	14	35.0
	16–20	6	15.0
	More than 20	12	30.0
Main working language	English	26	65.0
in addition to	Japanese or Korean	11	27.5
Mandarin Chinese	Other European languages	2	5.0
	Other Southeast Asian languages	1	2.5

TABLE 2.4 Profile of interviewees

Participant	Age group	Years of experience	Language pair
1	26–35	6–10	Mandarin Chinese/English
2	36–45	11–15	Mandarin Chinese/English
3	46–55	16–20	Mandarin Chinese/English
4	36–45	11–15	Mandarin Chinese/English
5	36–45	6–10	Mandarin Chinese/English
6	46–55	16–20	Mandarin Chinese/French
7	46–55	More than 20	Mandarin Chinese/English
8	36–45	11–15	Mandarin Chinese/English
9	36–45	11–15	Mandarin Chinese/English
10	46–55	11–15	Mandarin Chinese/English
11	46–55	More than 20	Mandarin Chinese/English
12	46–55	16–20	Mandarin Chinese/English
13	36–45	11–15	Mandarin Chinese/English
14	36–45	11–15	Mandarin Chinese/English
15	46–55	More than 20	Mandarin Chinese/Japanese

2.4 Questionnaire results

2.4.1 Workload

Table 2.5 provides detailed results of participants' workload, while Figure 2.1 presents a radar chart illustrating the weighted averages of interpreters' workload across three delivery modes (On-site, DI and Hybrid) and three periods (Pre-COVID, During COVID and Post-COVID). This radar chart visualization

TABLE 2.5 Workload across three periods

	Annual average before February 2020		Total number of days of work from March 2020 to October 2022 (32 months)		Total number of days of work from November 2022 to March 2024 (17 months)	
	Days	N	Days	N	Days	N
On-site	<20	4	<40	28	<30	9
	21–40	5	41–80	4	31–60	10
	41–60	6	81–120	4	61–90	8
	61–80	8	121–160	3	91–120	4
	81–100	6	161–200	0	121–150	5
	>100	11	>200	1	>150	4
	Weighted average	4	Weighted average	1.6	Weighted average	2.9
Distance	<20	37	<40	15	<30	29
	21–40	1	41–80	6	31–60	6
	41–60	0	81–120	5	61–90	3
	61–80	0	121–160	10	91–120	0
	81–100	2	161–200	1	121–150	1
	>100	0	>200	3	>150	1
	Weighted average	1.2	Weighted average	2.6	Weighted average	1.5
Hybrid	<20	36	<40	26	<30	26
	21–40	2	41–80	8	31–60	10
	41–60	0	81–120	2	61–90	4
	61–80	0	121–160	3	91–120	0
	81–100	0	161–200	0	121–150	0
	>100	2	>200	1	>150	0
	Weighted average	1.3	Weighted average	1.6	Weighted average	1.4

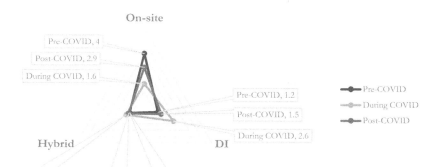

FIGURE 2.1 Radar chart of workload.

offers a comprehensive comparison of workload distribution trends, revealing patterns that may be less apparent in tabular data. On-site interpreting demonstrates a clear trend: it was most prevalent pre-COVID (weighted average 4.0), decreased significantly during COVID (1.6) and then rebounded post-COVID (2.9), though not reaching pre-pandemic levels. DI shows an inverse pattern, starting low pre-COVID (1.2), peaking during COVID (2.6) and slightly decreasing post-COVID (1.5), indicating its increased adoption during the pandemic. Hybrid interpreting remained relatively stable across all periods (pre-COVID: 1.3, during COVID: 1.6, post-COVID: 1.4), suggesting consistent but lower utilization compared to the other modes. This visualization effectively highlights the pandemic's impact on interpreting modes, showing a shift from on-site to DI during COVID-19, with a partial return to pre-pandemic patterns in the post-COVID period.

As a follow-up question to workload, the survey examined how frequently interpreters encountered three post-pandemic scenarios for clients who switched from on-site to DI during the pandemic. Among the 40 respondents, the most common scenario was clients reverting to on-site interpreting, with 23 (57.5%) reporting this occurring "very often" and 14 (35%) "sometimes", indicating a strong trend of returning to pre-pandemic practices. The scenario of clients continuing with DI was less frequent but still significant, with 6 (15%) encountering this "very often" and 18 (45%) "sometimes". Adopting hybrid interpreting was reported by 11 (27.5%) as occurring "very often" and 15 (37.5%) "sometimes". These results suggest that while the majority of clients reverted to on-site interpreting post-pandemic, a substantial portion either continued with DI or adopted hybrid models.

The survey also explored whether interpreters encountered a new demand for interpreting services due to the advantages of DI during the pandemic. Among the 40 respondents, 17 (42.5%) reported occasional encounters with such situations, while 11 (27.5%) were unsure. Only 2 (5%) frequently encountered this phenomenon, with 3 (7.5%) sometimes encountering it. Notably, 7 (17.5%) never experienced this situation. These findings suggest a moderate emergence of new clients during the pandemic, though the trend was not widely observed or recognized across the sample.

2.4.1.1 Working conditions

The following section examines the working conditions for DI. Regarding fee structures, most participants ($n = 33$, 82.5%) maintained the standard Taiwanese practice of using half-day rates as unit price. A minority used hourly ($n = 6$, 15%) or daily rates ($n = 1$, 2.5%). The majority ($n = 32$, 80%) charged equivalent fees for DI and on-site assignments, with few charging less ($n = 5$, 12.5%) or more ($n = 3$, 7.5%).

Contract modifications for DI assignments varied. While 19 respondents (47.5%) made no changes, an equal number of them modified disclaimer clauses. Other modifications included changes to working hours ($n = 7$, 17.5%), fees ($n = 6$,

15%), interpreters' rights and obligations ($n = 4$, 10%), clients' rights and obligations ($n = 3$, 7.5%) and team strength ($n = 1$, 2.5%).

Regarding work duration, 17 respondents (42.5%) reported that DI assignments
were "often" shorter than on-site jobs, 11 (27.5%) said "sometimes" and 4 (10%)
each reported "rarely", "always" or "never" experiencing shorter durations.

Team strength for DI largely mirrored on-site assignments ($n = 34$, 85%), with
some participants ($n = 5$, 12.5%) occasionally working alone. Only one participant
(2.5%) reported "often" having weaker team strength during DI.

Booth mate locations during DI varied, with most participants ($n = 26$, 65%)
ranking "separate locations" as most frequent. "Co-location at the client's designated place" was typically ranked second ($n = 28$, 70%), while "co-location at
interpreters' agreed place" was usually ranked last ($n = 26$, 65%).

For DI platforms, 35 respondents (87.5%) ranked video conferencing tools (e.g.,
Zoom, Webex) as the most frequently used. Makeshift systems (e.g., combination of messaging apps) were typically ranked second by 25 (62.5%) participants.
Dedicated RSI platforms (e.g., Interprefy, Kudo, Interactio) were ranked last by 26
(65%) participants. This ranking pattern indicates a clear client preference for commonly available video conferencing tools over specialized interpreting platforms
or improvised solutions.

2.4.1.2 Perception of DI

Participants were asked to choose three advantages of DI that they most agree
with among eight options. "Reducing commute" ($n = 36$, 90%), "less hassle (no
need to dress up, socialize, etc.)" ($n = 28$, 70%) and "personalized workspace"
($n = 25$, 62.5%) were the top three chosen.[1] Additional advantages mentioned
included more time with children for parents and reduced carbon footprint due to
less travel and catering needs.

Participants then selected five disadvantages they most agree with from 13
options. The top five were "bad audio quality" ($n = 34$, 85%), "difficult to interact
with others" ($n = 29$, 72.5%), "lack technical support" ($n = 23$, 57.5%), "operating different platforms and applications is troublesome" ($n = 23$, 57.5%) and "bad
video quality" ($n = 22$, 55%).[2]

Most participants "never" ($n = 25$, 62.5%) or "rarely" ($n = 10$, 25%) declined
assignments because they were DI ones, while few "sometimes" ($n = 3$, 7.5%) or
"often" ($n = 2$, 5%) did so.

After nearly three years of DI experience, about half of the participants maintained their original perceptions: 11 (27.5%) remained neutral, 7 (17.5%) still
viewed DI negatively and 3 (7.5%) retained positive impressions. Among those
whose opinions changed, 4 (10%) developed slightly worse views, and 1 (2.5%)
had much worse opinions. Conversely, 12 (30%) developed slightly better opinions, and 2 (5%) became very favorable toward DI.

Despite this positive shift, 34 (85%) of participants still ranked on-site interpreting as their most preferred mode. Hybrid was ranked second by 23 (57.5%) and last by 16 (40%). DI was ranked least preferred by 21 (52.5%), second by 14 (35%) and first by only 5 (12.5%).

2.4.2 Interviews

DI gained widespread adoption during the pandemic; however, as conditions improved, most settings reverted to in-person formats. DI continues to have applicable scenarios, particularly in business meetings, training courses and internal communications. Nevertheless, it presents certain technological and interactive limitations. Interpreters have adopted a rational attitude toward DI, recognizing its advantages and disadvantages while striving to adapt to this new work modality. Interviewees believed that DI is likely to coexist with on-site interpreting in the future, offering options for clients with diverse needs while simultaneously presenting interpreters with new opportunities and challenges.

2.4.2.1 Workload pattern

Interviewees generally agreed with the quantitative results of the workload pattern. Prior to the pandemic, DI was negligible, with organizers only adopting hybrid modes when a few speakers could not attend physically. Approximately six months after border restrictions were implemented, DI began to emerge and became the dominant mode of interpreting. However, due to the effective pandemic management by the Taiwan government, many organizers gradually preferred a hybrid approach, wherein local speakers and audiences participated on-site while foreign attendees who could not enter the country joined online. To mitigate potential technical issues related to internet connectivity, organizers frequently requested that foreign speakers pre-record their presentations, which were then played back during the event. These remote participants would then join the live event online during subsequent panel discussions or question-and-answer sessions. Purely virtual meetings were limited to specific client types and occasions, such as internal corporate meetings, working-level discussions and interviews. Large-scale events like academic conferences or annual congresses rarely adopted a fully virtual format.

When the pandemic began, there was a period when no meetings were held as everyone waited to see what would happen, thinking it might be similar to SARS and normalcy would return by the second half of the year. However, by June or July, numerous remote meetings started to emerge. Interestingly, perhaps because the situation in Taiwan was not too severe, even for online meetings, staff from the organizing body and interpreters would still gather at the same venue, while the audience and speakers participated online.

(Participant 8)

Smaller meetings, those with around a dozen participants, especially if they were scattered around the world, tended to remain online. Examples include board meetings, working-level discussions, or internal trainings. For these types of meetings, we continued to work virtually.

(Participant 1)

A common characteristic of meetings that continued to be held online or in hybrid mode after the pandemic was that the clients were often from the private sector, typically multinational corporations. In contrast, public sector clients rarely actively sought online or hybrid meeting formats.

(Participant 13)

Following the conclusion of the pandemic, the majority of clients opted to transition back to on-site interpreting from DI for several reasons. Face-to-face interactions were perceived as more effective in fostering relationships and trust, particularly between business partners and government officials. Clients sought enhanced value experiences, recognizing that on-site meetings facilitated more profound exchanges and networking opportunities. Industries like education and training and direct sales, which heavily rely on the ambiance of physical events, found DI less effective. Clients generally believed that on-site meetings yielded superior results, allowing participants to immerse themselves in the live atmosphere. Additionally, concerns about network stability and audio quality in virtual sessions influenced the shift back to on-site interpreting. Government agencies and public sector entities exhibited a marked preference for face-to-face meetings, further contributing to this trend. These factors collectively drove the post-pandemic shift from DI to on-site interpreting, reflecting a widespread desire for more direct and immersive communication experiences.

The client from the direct sales industry had experienced issues with virtual or hybrid events and expressed grave concern about internet connectivity. As a result, they preferred to fly the interpreters to the venue rather than have them participate online.

(Participant 14)

Although RSI is very convenient, people still preferred face-to-face communication. They enjoyed sharing meals and networking in person. Additionally, some organizers wanted to invite high-profile or famous dignitaries to their events, as their physical presence in the room had a significantly different impact compared to their online participation.

(Participant 4)

Multiple clients have expressed that they feel significantly more assured when interpreters are physically present. They believe their messages can be conveyed much more effectively when they can actually see the interpreters. Consequently, even for meetings that could be conducted perfectly online, these clients still preferred to travel to Taiwan and see the interpreters in the booth.

(Participant 15)

Despite the general trend toward returning to on-site interpreting, some clients continued to utilize DI or hybrid interpreting services after the reopening of borders for several reasons. Cost considerations played a significant role, as DI or hybrid formats eliminated the need for international speakers' travel expenses. For routine work meetings, DI proved more time and energy-efficient. Some business negotiations opted for DI due to its time-sensitive nature, allowing for quicker arrangement and execution of meetings. DI also facilitated the participation of individuals unable to attend in person, ensuring inclusivity. Some clients had grown accustomed to the convenience of remote meetings and preferred to maintain this practice. DI was particularly suitable for short, small-scale discussion meetings and proved highly convenient for multinational corporations' internal communications. Regular events, such as board meetings, often retained the DI format. In certain circumstances, environmental considerations influenced some clients' decisions, as online meetings were perceived as more eco-friendly.

> The client discovered that their budget could only accommodate one on-site conference per year if they had to fly several foreign speakers to the location. However, by holding events virtually, they could organize up to five events and even incorporate live streaming.
>
> *(Participant 11)*

> The news agency discovered that their media exposure remained consistent even when interviews were conducted online. Consequently, they continued to use remote interpreting services.
>
> *(Participant 9)*

> The reason why DI and hybrid modes still account for approximately 50% of my work is that these methods serve as effective alternatives for my clients. In recent years, large corporations have been required to report their carbon footprint, necessitating a reduction in travel. Additionally, some individuals might insist on participating online solely to reduce their own carbon emissions.
>
> *(Participant 13)*

2.4.2.2 Working conditions

The widespread adoption of general-purpose meeting platforms such as Zoom and Webex for remote interpreting can be attributed to several factors. These platforms have a low barrier to entry and are familiar to a broad audience, making them accessible to both interpreters and clients. Their functionality is typically sufficient to meet standard interpreting requirements, eliminating the need for specialized solutions in many cases. In contrast, professional RSI platforms often come with higher learning curves and increased costs, which can be deterrents for some users. Many clients have already invested in and become accustomed to these general-purpose platforms, making them a natural choice for DI needs. Furthermore, these platforms have integrated interpreting features, offering convenience and ease of use. It is also worth noting that some clients may be unaware of the existence of specialized RSI platforms, further contributing to the prevalence of general-purpose meeting software in remote interpreting scenarios.

> The client stated that they had already invested in professional plans for Webex. Since the platform includes built-in interpretation functions, they decided to simply activate and utilize these features, especially as they were already familiar with Webex.
>
> *(Participant 8)*

> I frequently work for these RSI platforms, so I'm quite familiar with them. When I asked about the cost of this service on behalf of another client, I was shocked to discover it's incredibly expensive. I'd dare say most businesses wouldn't even consider paying for it. Unless you're a huge multinational corporation or international organization needing a hundred languages, these platforms are overkill. For most companies here, it's simply not worth the hassle or the price tag.
>
> *(Participant 7)*

Interpreters' experiences with professional RSI platforms have been mixed. While these platforms are specifically designed for DI, they present several challenges. Interpreters report that these platforms often have higher equipment requirements, potentially necessitating additional investments. The platforms typically offer fixed remuneration rates, leaving little room for negotiation, which some interpreters find restrictive. A notable concern is the platforms' emphasis on technical aspects over interpreting quality, potentially giving an edge to younger interpreters more adept with technology. Communication between interpreters and clients is often limited on these platforms, and collaborations with assigned interpreting partners are not always harmonious. Some interpreters have expressed dissatisfaction with the user experience, reporting increased anxiety during work sessions. Furthermore, a number of interpreters feel that the design of these platforms does not adequately address the specific needs of interpreting work. These factors collectively contribute to a complex and sometimes challenging experience for interpreters using professional RSI platforms.

> On one occasion, I was compelled to work on an RSI platform because the client had already purchased a time package with them. This situation forced me to go

through the considerable hassle of upgrading my devices to meet their technical requirements. Additionally, I had to spend hours completing paperwork to register as a 'vendor' and take their mandatory training courses. The whole process was time-consuming and inconvenient.

(Participant 11)

I often find myself negotiating with the platforms about boothmate candidates and contract terms, especially cancellation policies. If a client insists on being able to cancel my job within 24 hours without compensating me for my loss, I counter with my own condition. I tell them that under those terms, I should have the right to decline their confirmed offer within 24 hours if I receive a better-paying job, and they would need to find someone else.

(Participant 10)

Some platforms treat interpreters as disposable and show little regard for quality. It's glaringly obvious that many of the boothmates they assign to me haven't received proper training. Working with these untrained colleagues is extremely challenging, and the issues extend far beyond just managing handovers. The problems affect the entire interpreting process and overall quality. I've even heard rumors of these platforms hiring individuals willing to work for a mere fifth of my rate.

(Participant 5)

RSI platforms have given rise to a new cohort of interpreters: young professionals who are highly skilled with technology. These interpreters are quick to accept offers, adept at juggling multiple devices and screens, and are comfortable working with limited or no preparation material. I suspect this mindset might be related to the COVID pandemic. Perhaps they view everything as ephemeral, leading them to place less importance on building relationships with clients. Similarly, they may not see the value in belonging to a community of interpreters or fostering trust and collegiality with their peers.

(Participant 7)

Despite the specificities of DI, most interpreters have not significantly altered their contract terms. Others have incorporated additional clauses to address the unique aspects of remote work. A key addition to many contracts is a disclaimer clause, absolving interpreters of responsibility for service interruptions due to network issues beyond their control. Some interviewees now explicitly delineate in the contract the technical responsibilities of the client, including provisions related to audio quality standards. Work hour definitions have also become more flexible, with some interpreters willing to adopt hourly billing in appropriate circumstances. However, for work during nighttime hours due to time zone differences, special surcharges may be applied. A notable addition to some contracts is the inclusion of terms regarding technical rehearsals or "sound check sessions". Some interviewees now stipulate that the first 30 minutes of technical testing are provided free of charge, with any additional time billed at overtime rates.

> Initially, my contractual terms remained unchanged. However, as I frequently found myself working during the wee hours due to time zone differences, I negotiated with clients to introduce a 'night-time fee'. For certain assignments, I had to book a co-working space or hotel room to avoid disturbing my family. In these cases, I would transfer the additional costs to the client.
>
> *(Participant 9)*

> For shorter assignments, typically lasting 40 to 60 minutes, I'm willing to work with hourly rates. This flexibility acknowledges the time saved on commuting, allowing me to easily accommodate another job in the same morning or afternoon. However, it's important to note that my hourly rate isn't simply my half-day rate divided by three. Instead, I maintain a base rate, which is approximately 70% of my half-day rate, to ensure fair compensation.
>
> *(Participant 10)*

The preparation process for DI requires interpreters to adapt their methods and equipment, with a crucial emphasis on testing equipment and network connections to ensure stable performance. Some interpreters go as far as taking screenshots of their hardware specifications and recording their computer screens during internet speed tests to provide evidence of their technical compliance. Many prepare backup devices to mitigate potential issues, with some even recording audio and video while the meeting is in session as a safeguard against quality disputes. The digital nature of DI has led to varied approaches in managing documents and input information: while some interpreters print more materials due to limited screen space, others invest in larger or multi-screen setups. A common strategy involves using multiple devices: the main screen for the meeting platform, a secondary screen for presentation documents and additional devices like tablets or smartphones for communication with interpreting partners. Interestingly, some interviewees have started to rely on the live captioning function on video conferencing platforms as it can assist their work when the sound quality is poor.

> With on-site interpreting, I've never had to worry about equipment or spend time reminding clients about technical issues, as that's the responsibility of the equipment vendor. However, DI has shifted much of this burden onto interpreters. I find myself repeatedly reminding clients about microphones, internet speed, camera angles, and other technical considerations. This additional responsibility is not only time-consuming but also mentally draining.
>
> *(Participant 6)*

Before each meeting, I run internet speed testing software while simultaneously recording my screen. I then send this recording to the clients, demonstrating that I'm fully prepared for the session. If I have a bad feeling about a particular assignment, I even go a step further by recording the entire session. This serves as proof if the audio quality becomes so poor that I have to stop interpreting. I do this because sometimes clients are skeptical of interpreters' claims about technical issues as they are often too busy to be aware of what is happening online. This proactive approach serves as comprehensive evidence of my technical readiness, ensuring that I cannot be held responsible for any connectivity issues that might occur during the meeting.

(Participant 11)

Since Zoom introduced live captioning, I've started requesting clients to activate this function as it serves as a valuable safety net. In cases where audio quality is subpar, live captioning can be a lifesaver. This is particularly crucial in DI settings, where it's challenging for colleagues to provide instantaneous assistance as they would in a physical booth. I've found that DI requires a heightened sense of self-reliance and preparedness.

(Participant 12)

2.4.2.3 Perception of DI

Interviewees' perspectives on the advantages and disadvantages of DI have remained largely consistent with past literature. There is a general acceptance of DI as an established work modality, rather than resistance to it.

The advantages of remote interpreting are numerous. The convenience of working without commuting saves significant time and energy. It opens up opportunities to work across different time zones, potentially increasing the volume and diversity of assignments. Clients often find it easier to arrange meetings, and interpreters can take on multiple assignments in a single day due to the elimination of travel time. This flexibility allows for better work-life balance and reduces stress related to appearance and attire, which can be especially attractive for interpreters who dislike socializing with others.

The most significant advantage of DI is undoubtedly the elimination of commute time and expenses. This not only increases my profit margin but also allows me to take on more assignments in the same amount of time. Additionally, it substantially reduces my carbon footprint, which is an important consideration. Perhaps most valuably, I get to spend more time with my family.

(Participant 11)

Some organizations have recognized the advantages of using Zoom's SI function and have consequently decided to hire interpreters directly. I've personally benefited from this trend, as it has increased my opportunities to work with a more diverse range of clients.

(Participant 3)

I don't like to socialize with people, and DI perfectly accommodates this preference. I can work from home, significantly reducing the need for in-person interactions with clients and colleagues.

(Participant 13)

However, DI is not without its challenges. The most significant concern is audio quality issues, which can directly impact the quality of interpretation. The lack of in-person interaction and atmosphere can affect the overall experience and potentially the interpreters' performance. Technical problems introduce additional stress, with interpreters often having to take on multiple roles to prevent and resolve technical issues. Some interviewees see the blurring of work-life boundaries as a notable drawback. Physical fatigue, particularly eye and ear strain, can be more pronounced in remote settings.

Despite all the technological advancements, I still believe that sound quality remains the most critical issue. Without good audio quality, it's impossible to deliver quality interpretation. I often have severe headaches for two to three days after working with bad sound.

(Participant 4)

While DI offers considerable time flexibility, allowing me to manage household tasks between assignments, I've found it has a significant downside. Work becomes deeply ingrained in daily life, blurring the boundaries between professional and personal time. The traditional rituals that once clearly delineated work from home life - such as dressing up for work, commuting to the venue, or treating myself to something special after a job - have disappeared. During the pandemic, I found myself missing these small but meaningful transitions.

(Participant 9)

Interpreters' perceptions of DI's impact on their role and status reveal a complex landscape dominated by the challenge of reduced physical presence. This lack of presence emerges as a central concern, with many interpreters feeling "invisible" in remote settings, struggling to effectively demonstrate their professional value when reduced to merely a voice in the system. This diminished visibility not only affects interpreters' sense of professional identity but also alters the dynamics of client relationships, creating an increased sense of distance.

Despite these challenges, some interpreters have found opportunities in the shift to DI. They report an expanded role that incorporates increased responsibility for technical aspects beyond pure linguistic interpretation, leading to a sense of added value through technical expertise.

> I've found that being physically present in the room significantly enhances my performance as an interpreter. When I'm there in person, I can feel the dynamics of the environment, which allows me to deliver a more refined and effective interpretation. The physical presence doesn't just improve my understanding of the context; it actually motivates me to engage more deeply and work harder. This tangible connection to the speakers, audience, and overall atmosphere creates a sense of immediacy and importance that's difficult to replicate in the DI mode. The energy of the room, the non-verbal cues, and the real-time feedback all contribute to a level of engagement that pushes me to elevate my performance.
>
> *(Participant 14)*

> Maintaining strong relationships with clients has become more challenging in the context of DI. The reduced face-to-face interaction inevitably leads to a gradual drifting apart.
>
> *(Participant 10)*

> I found myself taking on an additional role as a meeting consultant for my clients. I shared comprehensive advice on the pros and cons of online and hybrid meetings, guided them on how to arrange for better interpreting services, and even recommended specific devices and brands to purchase. This expanded role meant that clients came to rely on me not just for interpretation, but also for technical expertise. Interestingly, this consultative process became a powerful tool for strengthening client relationships. By providing this additional value, I was able to foster a deeper bond with clients, which proved instrumental in retaining their business and sustaining our professional relationships over time.
>
> *(Participant 15)*

Looking to the future, most interpreters envision DI coexisting with, rather than replacing, on-site interpreting. While generally maintaining a positive attitude toward technological advancements, concerns persist about the potential for AI to replace human interpreters in remote settings, as it is easier for RSI platform providers to introduce AI-powered interpreting in an online environment. Interpreters emphasize the importance of client education regarding the appropriate contexts for remote versus on-site interpreting.

DI has the potential to significantly transform the ecosystem of smaller language markets. With DI, clients are no longer constrained by geographical limitations and don't have to settle for mediocre interpreters in local markets. Instead, they

can now access top-tier talent from other regions or countries, potentially raising the overall quality of interpretation services available to them. However, it's
crucial to note that for this potential to be fully realized, connectivity issues need
to be effectively addressed first.

(Participant 6)

I think DI is here to stay, and given that DI is conducted on software platforms,
I'm convinced that our interpretations are being listened to by machines, likely
feeding into the development of AI systems. This leads me to believe that AI
interpreters may be introduced sooner than we anticipate. In light of this, I think
our focus should shift from concerns about DI itself to the more pressing issue
of AI in interpretation. The potential impact of AI on our profession could be
far more significant and transformative than the changes brought about by DI.

(Participant 3)

I believe that as technology continues to improve, we're moving towards finding an optimal balance between DI and on-site interpreting. The key factor is
the quality of audio and video; as these improve to meet our professional needs,
DI becomes an increasingly viable option for certain types of meetings. Our role
as interpreters is evolving to include advising clients on the best mode of interpreting service for their specific situations. By working collaboratively with our
clients in this way, we can create win-win situations that leverage the strengths
of both DI and on-site interpreting.

(Participant 8)

2.5 Discussion

Most of the findings of the present study largely align with recent research (Buján
and Collard 2022; Kincal and Ekici 2020; Mahyub Rayaa and Martin 2022; Matsushita 2022; Seresi and Láncos 2022). Compared with the results in Fan (2022),
interpreters in Taiwan still show a prevailing preference for on-site interpreting,
with 68% favoring traditional simultaneous interpreting in Fan (2022) and 85%
ranking it as their top choice in the current study. Audio quality concerns continue
to be a primary issue, identified by 85% of respondents in Fan (2022), and remaining a significant drawback in the latest findings. Both studies consistently highlight challenges in stakeholder interactions and recognize DI's advantages, such as
reduced commute time and increased flexibility. However, the recent study demonstrates a more nuanced understanding of DI among interpreters, reflecting their
increased experience and adaptation to this modality. This is evidenced by more
sophisticated technical preparations, contract modifications and an expanded role
that often includes technical consulting. These developments are often measures to
manage the credibility and communicative risks (Pym 2020) associated with much

more uncertain working conditions. The emergence of hybrid interpreting models and a more detailed analysis of workload patterns across different pandemic phases are also notable developments. Furthermore, the present study indicates heightened concerns about AI potentially replacing human interpreters in remote settings. These findings collectively suggest that while core perceptions of DI remain relatively stable, interpreters' approaches to and understanding of DI have evolved, reflecting both increased familiarity with the technology and a deeper consideration of its long-term implications for the profession.

The rise of the hybrid mode in Taiwan is intrinsically linked to the country's unique pandemic response and its distinctive interpreting service landscape. Taiwan's successful management of the COVID-19 pandemic, evidenced by the implementation of a "soft lockdown" only in May 2021 (Ministry of Health and Welfare 2021) for brief periods of time, allowed for the continuation of domestic activities while adhering to border restrictions. This scenario facilitated the widespread adoption of hybrid meetings, particularly among public institutions, which are significant employers of interpreting services in Taiwan. The public sector, often directly employing interpreters, provided opportunities for them to act as consultants during the pandemic. Similarly, some private companies, leveraging pre-existing close relationships with interpreters, involved them in the decision-making process for optimal arrangements and technological setups. This direct engagement between interpreters and clients resulted in a more nuanced understanding of interpreters' needs and challenges by the latter party. Consequently, interpreters in Taiwan often experienced relatively positive interactions with DI and hybrid modes, as they could effectively communicate their requirements.

However, this favorable scenario was not universal. When intermediaries were involved, the experiences varied significantly. Supportive agencies would negotiate and coordinate with the best interests of both clients and interpreters in mind, ensuring fair working conditions. Conversely, less scrupulous intermediaries would prioritize striking deals with end clients, often disregarding interpreters' rights and professional needs. A troubling trend that emerged during the pandemic was the "Uberization" of interpreting services (Fırat 2020; Giustini 2024; Özkaya Marangoz 2023), where interpreters were increasingly treated as on-demand workers. This shift toward a gig economy model undermined interpreters' efforts to optimize their performance. When interpreters become "gig workers", their presence in the value chain reduces. This potentially explains why the concept of "presence" (Moser-Mercer 2005) resurfaced as a crucial factor in shaping interpreters' perceptions of DI and hybrid interpreting. In contexts where interpreters had direct involvement in the planning and execution of remote or hybrid sessions, they reported a stronger sense of presence – a feeling of "occupying space" in the communicative event despite physical distance. This enhanced presence, including the co-location of colleagues, was associated with greater job satisfaction and perceived performance quality, which echoes the findings in past studies (Baumann 2023; Cheung 2022a, 2022b; Chmiel and Spinolo 2022; Seresi and Láncos

2022). However, the introduction of RSI platforms has led to the automation of numerous procedural aspects, potentially diminishing interpreters' sense of presence. These platforms often employ algorithms to manage various tasks, including interpreter assignment, boothmate matching, document distribution, pre-meeting communication, invoicing and troubleshooting (Giustini 2024). This automation, while potentially efficient, frequently results in minimal direct interaction between project managers and interpreters. Consequently, interpreters may feel excluded from the process, leading to a reduced sense of presence in the interpreting ecosystem. Furthermore, technological challenges, such as complex handover procedures, stringent hardware specifications and high internet speed requirements, often impeded interpreters' ability to perform optimally and maintain this sense of presence. Administrative barriers, including the pressure to accept assignments without adequate information and the inability to contact boothmates, further upended interpreters' established workflow and diminished their sense of involvement in the interpreting environment.

This dichotomy in experiences – between direct client interactions that enhanced interpreters' professional standing and sense of presence, and intermediary-mediated work that risked devaluing their services and reducing their presence – highlights the complex and often contradictory impacts of the pandemic on the interpreting profession in Taiwan. It underscores the need for a nuanced understanding of how different work arrangements, client relationships and the sense of presence shape interpreters' experiences in an increasingly digitalized work environment. These findings suggest that the effectiveness and acceptance of DI and hybrid interpreting may depend not only on technological solutions but also on the degree to which interpreters are integrated into the planning and execution of these new interpreting modalities, thereby enhancing their sense of presence and professional value.

The increasing digitalization of interpreters' work environment has led to a notable polarization of skills, as observed by several interviewees. This dichotomy manifests in two distinct areas of expertise: technological proficiency and traditional interpreting skills. Interpreters adept at managing multiple devices and platforms efficiently appear to be beneficiaries of this development, reflecting the tech-driven nature of modern interpreting environments. However, the ability to convey nuanced communication in person remains highly valued, particularly in settings where face-to-face interaction is preferred. This polarization creates a professional tension, requiring interpreters to balance technological competence with core interpreting skills. Interestingly, without prompting, several interviewees expressed concerns about the potential threat of AI in interpreting. For example, while some interpreters found features like live captioning beneficial, they also cautioned against over-reliance on such technologies. As one interviewee pointed out, "The progression from live captioning to translated live captioning is a small step that could potentially render human interpreting obsolete" (Participant 14). Indeed, the integration of AI into video conferencing and RSI platforms could significantly boost the advantage of interpreters who are tech-savvy but less proficient in traditional interpreting skills.

The findings of this study, when viewed through the lens of recent literature on the digitalization and platformization of interpreting work, reveal a complex landscape where interpreters grapple with technological changes that both enhance and potentially undermine their professional standing. Cronin and Delgado Luchner (2021) discussed the concept of "cognitive capitalism", where communication labor like interpreting is exploited through technologically advanced systems. This aligns with interviewees' sentiments regarding RSI platforms, where algorithms and technical criteria often take precedence over traditional interpreting skills. As some of our interviewees noted, these platforms seem to prioritize tech-savviness and responsive technical setups over the quality of interpretation itself. This shift in focus represents a form of de-skilling, where the core competencies of interpreting are potentially devalued in favor of technological proficiency, a phenomenon already mentioned in Fan (2022). Giustini (2024) further elaborated on this theme, describing how digital labor platforms are reshaping the interpreting industry. The concept of "real subsumption" that Giustini discussed is particularly relevant to our findings. Some interpreters in our study appear to be willingly subsumed into the gig economy model, despite their misgivings about DI and RSI platforms. This willingness to adapt, even in the face of recognized disadvantages, reflects the "transitional subsumption" Giustini described, where self-employed workers experience a lowering of independence as technology dictates specific labor conditions.

Similarly, despite complaints about the disadvantages of DI and the limitations of RSI platforms, many interpreters continue to accept these assignments, even post-pandemic. This behavior aligns with Giustini's (2024) observations about the "cash nexus", or the reduction of complex human relationships to financial transactions, in platform work, where interpreters may feel compelled to accept available work to maintain their income and professional relevance. The apparent downplaying of the potential impacts of DI on auditory and mental health is particularly concerning. Although most interviewees reported suspicions of hearing deterioration and mentioned increased fatigue and anxiety, they rarely decline job offers. It suggests that the pressure to remain competitive in a rapidly changing market may be leading interpreters to prioritize short-term employment opportunities over long-term well-being. This trend echoes Cronin and Delgado Luchner's (2021) concerns about the intensification of interpreter exploitation in the digital age.

The interplay between technological advancements and traditional interpreting skills observed in this study reflects a broader transformation within the profession. As interpreters in Taiwan and beyond continue to adapt to evolving work environments, the concept of "presence" in digital spaces emerges as a critical factor in shaping professional identity and practice. This shift necessitates a reconceptualization of what it means to be an interpreter in the digital age, challenging long-held notions of physical presence and interpersonal dynamics. The experiences documented here underscore the need for a more granular understanding of interpreter agency and professional value in increasingly digitalized and platformized work contexts. As the field continues to evolve, these insights from Taiwan's interpreting market may offer valuable perspectives on navigating the complex intersection of technology, professionalism and interpreter well-being.

2.6 Conclusion

This study analyzed the evolution of the perception of DI among Taiwan's interpreters, particularly in the context of the post-COVID era. The findings indicate a shift in how DI is viewed and utilized within the profession, influenced heavily by the COVID-19 pandemic. Initially, DI was minimally used before the pandemic, with a majority of interpreting conducted on-site. However, the pandemic catalyzed a rapid and substantial shift toward DI, with many interpreters adapting to this mode out of necessity. Post-pandemic, while there has been a significant return to on-site interpreting, DI has maintained a meaningful presence, particularly in specific sectors such as business meetings, training sessions and internal communications. Interpreters have generally adapted well to DI, recognizing its advantages, such as reduced commute times, increased flexibility and the ability to handle more assignments. However, challenges remain, particularly regarding audio quality, technical issues and the lack of personal interaction. These challenges have highlighted the need for robust technical preparations and adaptations in work habits. The perception of DI among interpreters has evolved to become more nuanced. While there remains a strong preference for on-site interpreting due to its interactive and immersive nature, interpreters have acknowledged the practicality and necessity of DI in certain contexts. This pragmatic acceptance suggests that DI is likely to remain a complementary mode alongside traditional on-site interpreting.

This study has several limitations that should be addressed in future research. The sample size of 40 survey respondents and 15 interviewees limits the generalizability of the findings. Additionally, the study's geographic specificity to Taiwan, which had a unique pandemic response, may not reflect DI adoption patterns in other countries or regions. The short-term nature of the observations also limits our understanding of long-term impacts on interpreters' careers and health. As technology and work practices continue to evolve, the perceptions and usage of DI may further change, necessitating ongoing research to capture these dynamics.

Future research could explore the long-term impacts of technological advancements on the interpreting profession. Incremental technological evolutions, such as improvements in delivery modes and the introduction of live translated captioning, have the potential to significantly alter the landscape of interpreting services. While these changes could enhance the efficiency and accessibility of interpreting, they also pose risks such as de-skilling and increased competition from AI-powered solutions. Researchers could examine how interpreters can best integrate these new tools while maintaining the core competencies of their profession. This investigation could lead to the development of best practices for incorporating new technologies in a way that supports interpreters rather than replacing them. Moreover, understanding the client perspective on DI and on-site interpreting will be crucial. Future studies could delve into how client preferences and requirements are shaping the demand for different interpreting modes and what this means for the future of the profession. Such research could include examining the cost-benefit analysis clients perform when choosing between on-site and DI services, particularly in terms of effectiveness, convenience and cost. Additionally, the long-term effects of increased screen time, remote work, subpar audio quality and cognitive load on interpreters' physical and mental health warrant further investigation. These factors, which have become more prevalent with the rise of DI, may have significant implications for the sustainability of the profession and the well-being of its practitioners.

In conclusion, the evolution of DI among Taiwan's interpreters reflects broader trends in the global interpreting profession. While challenges remain, the potential for technological advancements to enhance the profession is significant, provided that these changes are managed thoughtfully and inclusively. Future research should continue to monitor these trends, providing insights that help shape the future of interpreting in a rapidly changing technological landscape.

Notes

1 The remaining five advantages are: "increase job opportunities" ($n = 15$, 37.5%), "shorter duration of work" ($n = 7$, 17.5%), "increase income" ($n = 5$, 12.5%), "more audience listening to interpreting" ($n = 2$, 5%) and "safeguarding health" ($n = 2$, 5%).
2 The remaining eight disadvantages are: "difficult to handover" ($n = 19$, 47.5%), "affects auditory health" ($n = 16$, 40%), "illegal livestreaming or recording of interpretation made easier" ($n = 14$, 35%), "AI-enabled functions made easier impacting future prospect of interpreting profession" ($n = 6$, 15%), "less presence" ($n = 5$, 12.5%), "bad working conditions" ($n = 5$, 12.5%), "affects mental health" ($n = 2$, 5%) and "loneliness" ($n = 2$, 5%).

References

Anthropic. 2024. "Claude (Version 3.5) [AI Language Model]." https://www.anthropic.com.
ATLAS.ti Scientific Software Development GmbH. 2023. "ATLAS.ti Mac (Version 24.1.1) [Qualitative Data Analysis Software]." https://atlasti.com.
Baumann, Antonia. 2023. "The Pandemic Booth: How Spatial Reconfigurations during the Pandemic Influence Cooperation and Communication Among Conference Interpreters." *Interpreting and Society: An Interdisciplinary Journal* 3(2): 150–68. https://doi.org/10.1177/27523810231178880.
Braun, Sabine. 2017. "What a Micro-Analytical Investigation of Additions and Expansions in Remote Interpreting Can Tell Us About Interpreters' Participation in a Shared Virtual Space." *Journal of Pragmatics* 107: 165–77. https://doi.org/10.1016/j.pragma.2016.09.011.
Buján, Marta, and Camille Collard. 2022. "Remote Simultaneous Interpreting and COVID-19: Conference Interpreters' Perspective." In *Translation and Interpreting in the Age of COVID-19*, edited by Kanglong Liu and Andrew K. F. Cheung, 133–50. Singapore: Springer Nature.

Cheung, Andrew K. F. 2022a. "COVID-19 and Interpreting." *INContext: Studies in Translation and Interculturalism* 2(2): 9–14. https://doi.org/10.54754/incontext.v2i2.26.

Cheung, Andrew K. F. 2022b. "Remote Simultaneous Interpreting from Home or Hub: Accuracy of Numbers from English into Mandarin Chinese." In *Translation and Interpreting in the Age of COVID-19*, edited by Kanglong Liu and Andrew K. F. Cheung, 121–35. Singapore: Springer Nature.

Chmiel, Agnieszka, and Nicoletta Spinolo. 2022. "Testing the Impact of Remote Interpreting Settings on Interpreter Experience and Performance: Methodological Challenges inside the Virtual Booth." *Translation, Cognition & Behavior* 5(2): 250–74. https://doi.org/10.1075/tcb.00068.chm.

Cronin, Michael, and Carmen D. Luchner. 2021. "Escaping the Invisibility Trap." *Interpreting and Society: An Interdisciplinary Journal* 1(1): 91–101. https://doi.org/10.1177/27523810211033684.

Donovan, Clare. 2023. "The Consequences of Fully Remote Interpretation on Interpreter Interaction and Cooperation: A Threat to Professional Cohesion?" *INContext: Studies in Translation and Interculturalism* 3(1): 24–48. https://doi.org/10.54754/incontext.v3i1.59.

Fan, Chiaming D. 2022. "Remote Simultaneous Interpreting: Exploring Experiences and Opinions of Conference Interpreters in Taiwan." *Compilation and Translation Review* 15(2): 159–98. https://ctr.naer.edu.tw/v15.2/ctr150205.pdf.

Fırat, Gökhan. 2020. "Interpreting as a Service: The Uberization of Interpreting Services." *Journal of Internationalization and Localization* 8(1): 48–75. https://doi.org/10.1075/jial.20006.fir.

Frittella, Francesca M., and Susana Rodríguez. 2022. "Putting SmarTerp to Test: A Tool for the Challenges of Remote Interpreting." *INContext: Studies in Translation and Interculturalism* 2(2): 137–66. https://doi.org/10.54754/incontext.v2i2.21.

Giustini, Deborah. 2024. "'You Can Book an Interpreter the Same Way You Order Your Uber': (Re)Interpreting Work and Digital Labour Platforms." *Perspectives* 32(3): 441–59. https://doi.org/10.1080/0907676X.2023.2298910.

Hale, Sandra, Jane Goodman-Delahunty, Natalie Martschuk, and Julie Lim. 2022. "Does Interpreter Location Make a Difference? A Study of Remote vs Face-to-Face Interpreting in Simulated Police Interviews." *Interpreting* 24(2): 221–53. https://doi.org/10.1075/intp.00077.hal.

Harpe, Spencer E. 2015. "How to Analyze Likert and Other Rating Scale Data." *Currents in Pharmacy Teaching and Learning* 7(6): 836–50. https://doi.org/10.1016/j.cptl.2015.08.001.

Hennein, Rachel, and Sarah Lowe. 2020. "A Hybrid Inductive-Abductive Analysis of Health Workers' Experiences and Wellbeing during the COVID-19 Pandemic in the United States." *PLoS One* 15(10): e0240646. https://doi.org/10.1371/journal.pone.0240646.

Hoyte-West, Anthony. 2022. "No Longer Elite? Observations on Conference Interpreting, COVID-19, and the Status of the Post-Pandemic Profession." *Orbis Linguarum* 20(1): 71–77. https://doi.org/10.37708/ezs.swu.bg.v20i1.9.

Kincal, Şeyda, and Enes Ekici. 2020. "Reception of Remote Interpreting in Turkey: A Pilot Study." *RumeliDE Dil ve Edebiyat Araştırmaları Dergisi* 21: 979–90. https://doi.org/10.29000/rumelide.843469.

Liu, Kanglong, and Andrew K. F. Cheung, eds. 2023. *Translation and Interpreting in the Age of COVID-19*. Singapore: Springer Nature. https://doi.org/10.1007/978-981-19-6680-4.

Matsushita, Kayo. 2022. "How Remote Interpreting Changed the Japanese Interpreting Industry: Findings from an Online Survey Conducted during the COVID-19 Pandemic."

INContext: Studies in Translation and Interculturalism 2(2): 167–185. https://doi. org/10.54754/incontext.v2i2.22.

Mahyub Rayaa, Bachir, and Anne Martin. 2022. "Remote Simultaneous Interpreting: Perceptions, Practices and Developments." *The Interpreters' Newsletter* 27: 21–42. https:// doi.org/10.13137/2421-714X/34390.

Matsushita, Kayo. 2022. "How Remote Interpreting Changed the Japanese Interpreting Industry: Findings from an Online Survey Conducted during the COVID-19 Pandemic." *INContext: Studies in Translation and Interculturalism* 2(2): 167–85. https://doi. org/10.54754/incontext.v2i2.22.

Ministry of Health and Welfare. 2021. "CECC Raises Epidemic Alert Level for Taipei City and New Taipei City to Level 3 and Strengthens National Restrictions and Measures, Effective from May 15 to May 28, in Response to Increasing Level of Community Transmission." Crucial Policies for Combating COVID-19. Last modified June 11, 2021. https://covid19.mohw.gov.tw/en/cp-4868-61352-206.html.

Moser-Mercer, Barbara. 2003. "Remote Interpreting: Assessment of Human Factors and Performance Parameters." *Communicate!* Summer 2003. https://aiic.org/document/ 516/AIICWebzine_Summer2003_3_MOSER-MERCER_Remote_interpreting_ Assessment_of_human_factors_and_performance_parameters_Original.pdf.

Moser-Mercer, Barbara. 2005. "Remote Interpreting: Issues of Multi-Sensory Integration in a Multilingual Task." *Meta* 50(2): 727–38. https://doi.org/10.7202/011014ar.

Mouzourakis, Panayotis. 2006. "Remote Interpreting: A Technical Perspective on Recent Experiments." *Interpreting* 8(1): 45–66. https://doi.org/10.1075/intp.8.1.04mou.

Özkaya, Esra M. 2023. "Interpreting as a Service: The Uberization of Interpreting Services." *Abant Çeviribilim Dergisi* 1(1): 55–63. https://dergipark.org.tr/en/download/ article-file/3358314.

Proudfoot, Kevin. 2022. "Inductive/Deductive Hybrid Thematic Analysis in Mixed Methods Research." *Journal of Mixed Methods Research* 17(3): 308–26. https://doi.org/ 10.1177/15586898221126816.

Pym, Anthony. 2020. "Translation, Risk Management, and Cognition." In *The Routledge Handbook of Translation and Cognition*, edited by Fabio Alves and Arnt Jakobsen, 445–58. New York: Routledge.

Roziner, Ilan, and Miriam Shlesinger. 2010. "Stress and Performance in Remote Interpreting." *Interpreting* 12(2): 123–42. https://doi.org/10.1075/intp.12.2.05roz.

Seeber, Kilian G., Laura Keller, Rhona Amos, and Sophie Hengl. 2019. "Attitudes towards Video Remote Conference Interpreting." *Interpreting* 21(2): 270–304. https://doi. org/10.1075/intp.00030.see.

Seresi, Marta, and Petra L. Láncos. 2022. "Teamwork in the Virtual Booth—Conference Interpreters' Experiences with RSI Platforms." In *Translation and Interpreting in the Age of COVID-19*, edited by Kanglong Liu and Andrew K. F. Cheung, 169–96. Singapore: Springer Nature.

Thompson, Jamie. 2022. "A Guide to Abductive Thematic Analysis." *The Qualitative Report* 27(5): 1410–21. https://doi.org/10.46743/2160-3715/2022.5340.

Zhu, Xuelian, and Vahid Aryadoust. 2022. "A Synthetic Review of Cognitive Load in Distance Interpreting: Toward an Explanatory Model." *Frontiers in Psychology* 13: Article 899718. https://doi.org/10.3389/fpsyg.2022.899718.

3

A SYSTEMATIC REVIEW OF TECHNOLOGY INTEGRATION IN INTERPRETER EDUCATION

Current applications, challenges, opportunities and regional variations

Wei Guo and Feng Cui

3.1 Introduction

The integration of technologies into interpreter education has long been a topic of discussion among researchers and practitioners alike. As the fourth industrial revolution gains momentum, driven primarily by advancements in Artificial Intelligence, particularly generative artificial intelligence (AI), technologies such as automatic speech recognition, deep learning and neural networks are revolutionizing the field of interpreting. These technological breakthroughs hold immense potential for enhancing the teaching and learning of interpreting skills and competencies. The evolution of interpreter education has witnessed a shift from traditional teacher-led classrooms to learner-centered and interactive paradigms (Chan 2023b). Adapted from definitions by scholars in the field of applied translation (Costa et al. 2014; Fantinuoli 2016; Zhao 2017; Braun 2019; Wang and Yang 2019; Wang and Li 2022), this article defines the concept of technologies in interpreter education as a variety of technologies applied by interpreting teachers or researchers to enhance the teaching and learning of interpreting skills and competencies.

From the introduction of interpreting booths and sound transmission systems in the 1920s to the emergence of setting-oriented platforms and intelligent process-oriented technologies in the 21st century, technology has continuously reshaped the way interpreters work and learn. Existing research on interpreting technology can be broadly categorized into two strands. The first strand focuses on descriptive studies that explore the latest technological advancements and how they are integrated into interpreting practices. Key areas of interest include computer-assisted interpreting (CAI), machine interpreting (MI) and remote interpreting (RI). CAI tools, for instance, assist interpreters in tasks such as terminology management, material organization and information extraction, while

DOI: 10.4324/9781003597711-4

MI technologies aim to automate the translation of spoken language, albeit with limitations. RI, on the other hand, enables interpreter-mediated communication across geographical boundaries, offering new opportunities for service delivery. The second strand of research employs empirical methods to investigate the actual effects of these technologies on the interpreting process and quality. Studies have demonstrated the positive influence of technologies like CAI tools in enhancing terminological precision and reducing omissions, while automatic speech recognition applications have been shown to assist interpreters in number rendition and reduce cognitive load.

Despite the progress made in integrating technologies into interpreting practices, their application in interpreter education remains limited. This is primarily due to challenges such as students' motivation and technological proficiency levels (Chan 2023b) and the availability of resources, provided by the educational institutions, including curriculum (Prandi 2020), technological equipment, software packages and financial budgets (Chan 2023b).

Given these circumstances, this systematic review aims to provide a comprehensive overview of the current applications of technology in interpreter education, including the technologies and tools adopted in computer-assisted interpreter training (CAIT). Specifically, this review will address three key research questions:

1 What are the current trends and applications of interpreting technology in educational settings?
2 What are the challenges and opportunities associated with integrating technology into interpreting teaching and learning?
3 How do regional and institutional differences influence the integration of technology in interpreter education?

By exploring these questions, this review aims to identify gaps in previous studies, categorize the literature under relevant themes and facilitate an in-depth exploration of the opportunities, challenges, regional variations and best practices in integrating technology into interpreter education. The findings of this review will have practical implications for interpreter teaching, educational management and the future of interpreter education in the digital era.

3.2 Research methods

This systematic literature review followed Xiao and Watson's (2019) eight-step process and adhered to the PRISMA (Preferred Reporting Items for Systematic Reviews and Meta-Analyses) guidelines (Moher et al. 2009). The eight steps include: (1) Formulating the research problem; (2) Developing and validating the review protocol; (3) Searching the literature; (4) Screening for inclusion; (5) Assessing quality; (6) Extracting data; (7) Analyzing and synthesizing data; and (8) Reporting findings, as shown in Figure 3.1.

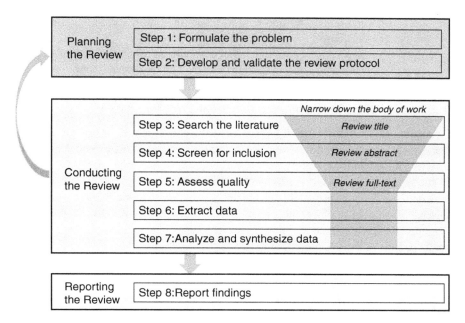

FIGURE 3.1   Process of systematic literature review (Xiao and Watson 2019, 11).

PRISMA serves as a widely adopted framework for conducting rigorous litera-ture reviews. The methodology provides a four-phase flow diagram and a check-list comprising 27 items to ensure comprehensive reporting of systematic reviews and meta-analyses (Moher et al. 2009). This checklist covers all necessary steps, ranging from title and abstract to introduction, methods, results, discussion and conclusions. The four-phase flow diagram consists of identification, screening, eli-gibility and inclusion. While PRISMA is instrumental in organizing the literature, the eight-step process by Xiao and Watson (2019) offers a detailed guideline for the search stages. In this review, we adhere to both frameworks to ensure a thorough and unbiased examination of the literature.

3.2.1   Formulating the research problems

The initial step is formulating the research problems, which is essential as "litera-ture reviews are research inquiries, and all research inquiries should be guided by research problems" (Xiao and Watson 2019). Following an initial scan of the litera-ture on technologies in interpreting pedagogy, the authors conducted a pre-review mapping to identify distinct subtopics. The following research questions were for-mulated based on the identified themes:

1 What are the current applications of technology integration in interpreter educa-tion, particularly in terms of the technologies and tools employed in CAIT?

2 What challenges and opportunities arise when integrating technology into interpreting teaching settings?

3 How do regional and institutional differences impact the integration of technology into interpreter education, in terms of both challenges and opportunities?

3.2.2 Developing and validating the review protocol

According to Xiao and Watson (2019), the review protocol is "a pre-set plan that specifies the methods utilized in conducting the review". This protocol is crucial for ensuring a quality literature review without bias. It explicitly states the purpose of the study, inclusion and exclusion criteria (see Table 3.1), screening procedures and data extraction strategies. Exclusion criteria were particularly important to guarantee the quality of articles included in the review (Metruk 2022). This rigorous approach ensures that the findings are grounded in a comprehensive and unbiased analysis of the available literature.

3.2.3 Searching the literature

Adhering strictly to the review protocol, the authors conducted a comprehensive literature search using two renowned databases in the social sciences: Scopus and Web of Science (WoS). Initially, in Web of Science and Scopus, the authors searched using the combined terms "interpret* pedagogy" AND "technology", resulting in only one result in both databases. To capture more relevant literature, the search terms were refined into three separate groups: "interpret*" AND "pedagogy" AND "technology", "interpret*" AND "teach*" AND "technology" and "interpret*" AND "train*" AND "technology". However, most results from these three groups were deemed irrelevant to the research topic. To enhance relevancy, the search scope was limited to "Educational research and linguistics" in WoS and "Social science and Art and humanities" in Scopus. This refinement yielded 2,150 results in WoS and 2,629 results in Scopus, totaling 3,424 unique results after duplicate checks in Microsoft Excel.

TABLE 3.1 Inclusion and exclusion criteria

Criterion	Inclusion	Exclusion
Types of literature/studies	All relevant quantitative and qualitative articles on the topic	Duplicated and unrelated literature; literature without clear authors; literature not written in English; conference proceedings
Timeframe for the topic of technologies in interpreting pedagogy	Before May 1, 2024	May 1, 2024

Additionally, the authors identified the top four journals related to interpreting and education on the current research topic: *Education and Information Technologies, Interpreter and Translator Trainer, Meta* and *Interpreting.* The three groups of terms were searched in these four journals, resulting in 401, 169, 138 and 124 results, respectively, after duplicate checks. Combining database searches and manual searches in these journals, a total of 4,256 articles were identified (as shown in Figure 3.2), with 4,224 articles remaining after removing duplicates. The final search in the references of the databases and manually searched articles was conducted on May 1, 2024.

3.2.4 Screening for inclusion

Following the initial screening of titles, articles unrelated to interpreting were removed, leaving 746 articles in the two databases. Subsequently, the authors reviewed titles and abstracts again to narrow down the literature scope by excluding studies focused on translation technologies and technologies used in interpreting practices. If inclusion decisions were difficult based on abstracts alone, the conclusion sections were consulted for final determination. This process resulted in 102 articles from the database search for detailed reading to further assess their relevance to the topic of integrating technologies in interpreting pedagogy. Unfortunately, 11 relevant articles were inaccessible for downloading, leaving 91 articles with full texts for further quality assessment.

Similarly, articles manually searched in the four journals underwent screening based on titles. In *Education and Information Technologies*, 323 search results predominantly focused on general curriculum designs, classroom management and improving teaching and students' performance, with 43 articles covering specific subjects ranging from primary to high schools, and 30 articles highlighting the use

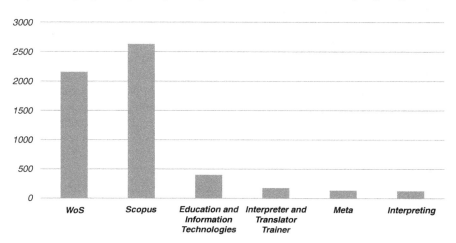

FIGURE 3.2   Count of articles in both databases and four top related journals.

of technologies in English as a Foreign Language (EFL) learning. Consequently, only five articles from Education and Information Technologies were retained for further analysis. For *Interpreter and Translator Trainer*, 145 results were excluded due to their irrelevance to the research topic, leaving 24 articles for full-text screening. Most studies in *Meta* were centered on translation technologies and interpreting practice settings, resulting in only 16 articles on the topic of technologies used in interpreting pedagogy for further screening. In *Interpreting*, studies predominantly focused on interpreting practice settings and emphasized the multidisciplinary framework of technology uses, leading to the inclusion of 35 articles for eligibility checks. In conclusion, a total of 80 articles were identified through initial screening across the four journals.

3.2.5 Assessing quality

Based on data screening, 91 articles from WOS and Scopus were obtained as full texts for quality assessment. Though quality assessment is not an essential step for some kinds of descriptive and critical reviews (Xiao and Watson 2019), it is necessary to evaluate the quality of articles to comprehend each study better to compare and integrate findings (Ludvigsen et al. 2016 as cited in Xiao and Watson 2019). The current review adopted the criterion of "internal quality" proposed by Petticrew and Roberts (2006, as cited in Xiao and Watson 2019) to prevent potential methodological biases. The authors appraised the studies according to a checklist approved by the reviewers and analyzed the literature from many aspects, including research focus, data collection and analysis methods, results and conclusions.

Of the 91 articles from the databases, many addressed interpreting technologies but did not focus on integrating them into interpreting pedagogy, leading to their exclusion. For example, a variety of studies focus on integrating technologies into interpreting practice, either about the use of digital pen in interpreting practice and research (Orlando 2014), prediction of simultaneous interpreting performance based on eye-tracking technology (Amos et al. 2023), tablet-based interpreting through interviews with practitioners (Goldsmith 2018) or the use of ASR-CAI tools to improve interpreter's performance. Other studies, though, investigated interpreter's mental and cognitive factors with the help of technologies, such as utilizing eye-tracking-based method to explore how training and experience affect student interpreter's cognitive load, interpreting speed and output quality (Su and Li 2021), using eye-tracking, heart rate and galvanic skin response to test interpreter's stress in COVID-19 medical scenario (Li et al. 2022). As a result, 37 articles were remained in database searching.

The manual searches from the journals created 80 full-text articles, 35 coming from *Interpreting*, 24 from *The Interpreter and Translator Trainer*, 16 from *Meta* and 5 from *Education and Information Technologies*. In the current stage, the researchers read the full texts of literature to decide whether they should be included based on the inclusion criterion. Only one out of five articles from

Education and Information Technologies focus on technology integration in interpreting pedagogy (Tian and Yang 2023), the remaining four were excluded for they either concerned translation technologies (Qassem and Al Thowaini 2024; Wang et al. 2024) or focused on English language learning (Hu and McGrath 2012; Yang 2024). Although all the studies from *The Interpreter and Translator Trainer, Interpreting* and *Meta* concerned interpreter training or interpreting technologies, not all articles highlighted both, either using new technologies to boost interpreting practice or research (Chen 2020; Petite 2005; Slettebakk Berge 2018) or underscoring assessment of interpreting based on new technologies but without further exploring pedagogical implications (Bernstein and Barbier 2000; Han 2021; Lu and Han 2022). In consequence, ten articles from *The Interpreter and Translator Trainer,* seven articles from *Interpreting* and five articles from *Meta* were included for further data extraction.

Resultantly, 60 articles from databases and journal searches were included for detailed analysis in this systematic literature review. During the phase of eligibility, some more relevant papers came out from the full texts included articles. After secondary searching and thorough reading of the literature, seven more articles were included for further analysis. Consequently, a total of 67 articles met the inclusion criteria and underwent thorough analysis. The entire screening process is depicted in Figure 3.3.

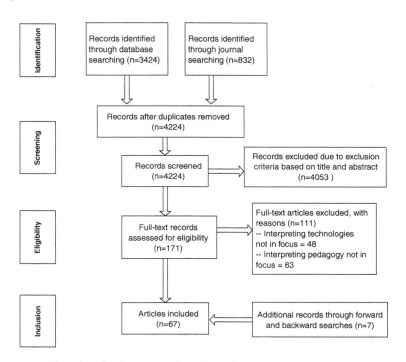

FIGURE 3.3   Flowchart for the systematic review of technologies in interpreting pedagogy.

3.2.6 Extracting data

Data extraction is a crucial step in the literature review process, as it involves coding the retrieved literature to facilitate in-depth analysis. As outlined by Saldaña (2016), thematic analysis involves the strategic assignment of codes to specific portions of data, ensuring that their essential characteristics are captured accurately. In this study, data coding was guided by the topics and subtopics of the research, spanning interpreting technologies, interpreting pedagogy, challenges and potential opportunities and regional differences. To streamline this process, the authors employed Nvivo, a qualitative data analysis software, to create and systematically organize these codes. After conducting full-text readings of all included articles and considering the study's topic, three primary themes were established (refer to Table 3.2). These themes were then manually refined and reorganized to enable deeper exploration of the data in line with the three research questions.

3.2.7 Analyzing and synthesizing the data

After meticulously coding all included studies through thematic analysis, the next step entails a rigorous process of analyzing and synthesizing the amassed data. This phase primarily involves descriptive analysis, where tables and figures are utilized to present the themes in a clear and concise manner. Through this analytical lens, the descriptive themes are refined into more profound analytic themes (Xiao and Watson 2019), effectively weaving them into a coherent narrative.

3.2.8 Reporting the findings

The culmination of a systematic review lies in reporting the findings pertaining to the identified topics and subtopics. This study embraces an integrative review approach, which accommodates the diversity of research designs, spanning quantitative, qualitative and mixed-methods studies (Shaffril et al. 2020). The focus will be on highlighting emergent themes and noteworthy findings, while briefly touching upon their implications for future research endeavors.

3.3 Results

3.3.1 Current applications of technologies in interpreter education

The systematic review encompassed 67 carefully selected articles, sourced from database searches and manual journal screenings conducted up to May 1, 2024. These studies were primarily retrieved from Web of Science and Scopus, with four key journals standing out in relevance: *Education and Information Technologies, Interpreter and Translator Trainer, Meta, and Interpreting*. The aim of this review is to provide a holistic overview of the current trends, gaps and regional focuses

TABLE 3.2 Themes in included articles

	Students		Teachers		Educational Institutions	
	Opportunities	Challenges	Opportunities	Challenges	Opportunities	Challenges
Aguirre Fernández Bravo (2020)	1	0	0	0	1	0
Ahrens et al. (2021)	1	0	0	1	0	3
Alcaide-Martínez and Taillefer (2022)	1	0	0	0	0	0
Arzık-Erzurumlu (2020)	2	0	0	0	0	0
Bale (2013)	1	0	0	0	0	0
Blasco Mayor et al. (2007)	3	0	2	1	0	1
Braun and Slater (2014)	2	0	0	0	0	0
Braun et al. (2020)	3	0	0	0	0	2
Castillo-Rodríguez et al. (2023)	2	1	0	1	0	1
Chan (2023a)	2	0	0	0	0	1
Chapman (1977)	0	0	0	0	0	1
Crezee et al. (2015)	2	0	0	0	0	0
D'Hayer (2012)	2	0	0	0	0	1
Eraslan et al. (2020)	2	0	0	0	0	0
Eser et al. (2020)	1	1	1	0	1	2
Gerbe et al. (2021)	3	0	1	0	0	0
Han et al. (2022)	2	0	1	0	0	0
Hansen and Shlesinger (2007)	3	0	0	0	0	1
Huang et al. (2023)	2	0	0	0	1	0
Hunt-Gómez and Moreno (2015)	1	0	1	0	1	0
Huỳnh and Nguyen (2019)	0	1	0	0	1	1
Ibrahim-González (2011)	2	0	0	1	0	2
Bernstein and Barbier (2000)	0	0	2	0	2	0
Jiang and Lu (2021)	2	0	0	0	0	0
Ko (2006)	0	0	0	0	0	2

(Continued)

TABLE 3.2 (Continued)

	Students		Teachers		Educational Institutions	
	Opportunities	*Challenges*	*Opportunities*	*Challenges*	*Opportunities*	*Challenges*
Ko (2008)	2	0	0	1	0	0
Ko and Chen (2011)	0	0	0	0	1	1
Lee and Huh (2018)	0	0	1	0	1	0
Lei and Zhu (2013)	2	0	0	2	1	0
Leitão (2020)	1	0	0	0	0	2
Li and Jiang (2021)	1	0	1	0	1	1
Lim (2013)	4	0	0	0	0	1
Lim (2020)	2	0	0	0	0	0
Rudvin et al. (2022)	4	0	0	0	1	0
Mankauskienė and Svetlana (2024)	1	1	0	0	2	0
Manko and Khitsenko (2019)	8	1	1	1	1	5
Mohammed and Mustafa (2023)	2	0	0	0	0	2
Moser-Mercer et al. (2005)	2	0	0	0	1	0
Nai and Hassan (2023)	1	0	0	0	0	0
Nai and Hassan (2022)	3	0	0	0	1	0
Orlando (2010)	2	0	0	0	0	0
Orlando (2015)	3	0	0	0	0	0
Orlando and Hlavac (2020)	1	0	0	0	0	0
Peng (2022)	3	0	0	0	0	0
Perramon and Ugarte (2020)	0	0	0	0	0	2
Yang (2022)	0	0	1	0	0	0
Postigo and Parrilla (2023)	2	0	0	0	0	0
Rodríguez Melchor et al. (2020)	3	0	0	0	0	1
Rodríguez Melchor (2020)	0	0	1	0	1	0
Ruiz Mezcua (2019)	0	1	0	0	0	2

Source						
Şahin (2013)	3	0	0	0	1	0
Sayaheen et al. (2023)	1	1	2	1	0	3
Schweda-Nicholson (1985)	4	0	0	0	0	0
Seeber and Luchner (2020)	2	0	0	0	0	0
Seresi and Horváth (2020)	1	0	0	0	0	1
Svetlana et al. (2020)	2	0	0	0	1	0
Tian (2014)	1	0	0	0	0	0
Valero Garcés (2020)	3	0	0	0	1	0
van Dyk (2010)	2	0	0	0	1	0
Wang et al (2023)	0	0	1	0	1	1
Wei et al. (2022)	1	0	0	0	0	0
Xiong (2021)	0	0	1	0	0	0
Xu (2018)	2	0	0	0	0	0
Ye (2020)	3	0	0	0	0	0
Yu and Van Heuven (2017)	0	0	0	0	2	0
Yuan et al. (2023)	2	0	0	0	0	0
Zhong et al. (2021)	1	0	1	0	1	0

in integrating various technologies into interpreter education, highlighting both widely adopted and underexplored technologies.

A notable trend in publication is the surge in recent research, with 49 of the 67 articles published within the past decade (2014–2023). This indicates a growing interest and development in the field of interpreter education technologies. The demographics of the literature are presented in Table 3.3, ordered by publication year.

Regarding technology integration trends, most studies (45 out of 67) focused on the application of interpreting devices. This category includes a range of technologies designed to facilitate the interpreting process, reflecting a significant interest in practical tools that can be directly implemented in interpreter training. These integrations encompass a variety of advancements, such as RI platforms that allow for flexible service delivery through video conferencing, speech recognition software aiding in real-time note-taking, mobile applications tailored for interpreters' needs and advanced simultaneous interpretation systems for conference settings. The integration of these devices into educational settings is highlighted by recent publications from China (e.g., Wang et al. 2023) and Spain (e.g., Castillo-Rodríguez et al. 2023), which signifies a trend toward leveraging technology to enhance interpreter training and education.

Information retrieval technologies, which assist interpreters in preparing by accessing and organizing relevant terminology and background information, were also a critical area of focus, with 12 articles discussing their use. The mentioned information retrieval technology tools are usually designed to assist interpreters in various ways, including but not limited to material organization, automated information extraction and instant access to and advanced search across multiple databases (Fantinuoli 2016; Prandi 2020). These technologies could be utilized to automatically extract relevant information from large volumes of text, such as conference proceedings, reports and articles. This saves time and ensures interpreters are equipped with comprehensive background information on the certain subject matter. Notable contributions come from the UK (e.g. Mankauskienė and Svetlana 2024) and China (e.g. Yang 2022), underscoring the global relevance of this technology.

Computer applications, including both general software and specialized tools for interpreting tasks, were covered in nine articles. These applications range from general computer software to more specialized tools designed for interpreting tasks. Computers and the internet have profoundly transformed the methods by which interpreters acquire knowledge and prepare for their tasks. MI technology such as Pixel Buds involves utilizing algorithms and linguistic databases to automatically convert spoken words from one language into another (Fantuoli 2018; Jekat 2015). Advanced MI systems may now incorporate AI techniques, such as neural networks, to improve accuracy and handle linguistic nuances. The programs or systems play a vital role in facilitating cross-linguistic communication and are widely used in various domains, including business, diplomacy and media. Research in this area is diverse, with contributions from countries like Malaysia (e.g. Yuan et al. 2023) and Turkey (e.g. Arzık-Erzurumlu 2020).

Finally, other innovative interpreting technologies, though less extensively covered, showed promising potential for enhancing interpreter education, as evidenced by nine articles dedicated to this category. These technologies, though not as widely covered as interpreting devices or information retrieval tools, show promising potential for enhancing interpreter education. Examples include research from Spain (e.g. Postigo and Parrilla 2023) and Australia (e.g. Orlando and Hlavac 2020). These findings underscore the importance of continued exploration and investment in diverse technological solutions to advance interpreter education and meet the evolving needs of the profession.

The review also revealed disparities in research attention across different technology categories.

Remote or online teaching technologies were the subject of 22 articles, with four specifically addressing remote teaching during the COVID-19 pandemic (Castillo-Rodríguez et al. 2023; Mohammed and Mustafa 2023; Han et al. 2022; Li and Jiang 2021). Additionally, nine articles explored the utilization of Virtual Reality in interpreter education. However, certain technologies received minimal research attention. For instance, the integration of AI and tablets into interpreting training was only addressed in one article each, while digital pen technology was discussed in only two studies (Orlando 2010, 2015).

Regarding other interpreting technologies, challenges primarily concern technological barriers, especially those affecting automatic speech recognition (ASR) systems. These barriers, including accents, pronunciation preferences, background music and breakdowns in fluency, as discussed by Huang et al. (2023) in their Chinese study, pose significant challenges. However, despite these challenges, other interpreting technologies offer both students and teachers opportunities. For students, they can lead to more accurate interpreting, enhanced engagement and improved learning outcomes, as demonstrated by Orlando and Hlavac (2020) in their Australian study. For teachers and universities, they facilitate more efficient testing and reduced costs, as noted by Bernstein and Barbier (2000) in their American study. However, ethical issues surrounding technology use, such as confidentiality and privacy concerns, as explored by Orlando (2015) in his Australian study, also need to be carefully considered.

The findings of this review shed light on prevalent trends and gaps in the integration of technology into interpreting pedagogy. Notably, while certain technologies like interpreting devices and online teaching platforms have received considerable attention, others such as the integration of virtual reality simulations or AI-assisted tools in interpreter education remain under-researched. Future studies should aim to explore these overlooked areas to ensure a comprehensive understanding of technology's role in enhancing interpreter training. This includes investigating the effectiveness of new technologies in improving interpreters' language proficiency, cultural understanding and practical interpreting skills, ensuring that interpreter education keeps pace with technological advancements for optimal training outcomes.

TABLE 3.3 Profile of included articles

Author(s)	Country/territory	Categories
Mankauskienė and Svetlana (2024)	Lithuania and UK	Information retrieval
Yuan et al. (2023)	Malaysia	Computer application
Wang et al. (2023)	China	Interpreting devices application
Castillo-Rodríguez et al. (2023)	Spain	Interpreting devices application
Mohammed and Mustafa (2023)	Iraq	Interpreting devices application
Sayaheen et al. (2023)	Jordan	Interpreting devices application
Huang et al. (2023)	China	Other interpreting technologies
Postigo Pinazo and Parrilla Gómez (2023)	Spain	Other interpreting technologies
Nai and Hassan (2023)	Malaysia	Other interpreting technologies
Peng (2022)	Taiwan	Other interpreting technologies
Yang (2022)	China	Information retrieval
Han et al. (2022)	China	Interpreting devices application
Wei et al.(2022)	China	Interpreting devices application
Rudvin et al. (2022)	Italy	Interpreting devices application
Nai and Hassan (2022)	China and Malaysia	Interpreting devices application
Chan (2023a)	China	Interpreting devices application
Alcaide-Martínez and Taillefer (2022)	Spain	Other interpreting technologies
Zhong et al. (2021)	China	Computer application
Xiong (2021)	China	Information retrieval
Ahrens et al. (2021)	German	Interpreting devices application
Gerbe et al. (2021)	Australia	Interpreting devices application
Li and Jiang (2021)	China	Interpreting devices application
Arzık-Erzurumlu (2020)	Turkey	Computer application
Lim (2020)	China	Information retrieval
Leitão (2020)	Europe	Information retrieval
Jiang and Lu (2021)	China	Interpreting devices application
Rodríguez Melchor (2020)	Spain	Interpreting devices application
Svetlana et al. (2020)	Europe	Interpreting devices application
Eraslan et al. (2020)	Turkey	Interpreting devices application
Seeber and Luchner (2020)	Europe	Interpreting devices application
Seresi and Horváth (2020)	Europe	Interpreting devices application
Aguirre Fernández Bravo (2020)	Spain	Interpreting devices application
Perramon and Ugarte (2020)	Canada	Interpreting devices application
Valero Garcés (2020)	Spain	Interpreting devices application
Rodríguez Melchor et al. (2020)	Europe	Interpreting devices application
Eser et al. (2020)	Malaysia and Australia	Interpreting devices application

(Continued)

TABLE 3.3 (Continued)

Author(s)	Country/territory	Categories
Braun et al. (2020)	UK	Interpreting devices application
Ye (2020)	China	Interpreting devices application
Orlando and Hlavac (2020)	Australia	Other interpreting technologies
Huỳnh and Nguyen (2019)	Vietnam	Interpreting devices application
Manko and Khitsenko (2019)	Ukraine	Interpreting devices application
Xu (2018)	China	Information retrieval
Ruiz Mezcua (2019)	Spain	Interpreting devices application
Lee and Huh (2018)	Korea	Interpreting devices application
Yu and Van Heuven (2017)	Chinese	Other interpreting technologies
Hunt-Gómez and Moreno (2015)	Spain	Information retrieval
Crezee et al. (2015)	New Zealand	Information retrieval
Orlando (2015)	Australia	Interpreting devices application
Tian (2014)	China	Interpreting devices application
Braun and Slater (2014)	UK	Interpreting devices application
Lei and Zhu (2013)	China	Information retrieval
Bale (2013)	UK	Information retrieval
Lim (2013)	China	Interpreting devices application
Şahin (2013)	Turkey	Interpreting devices application
D'Hayer (2012)	UK	Interpreting devices application
Ko and Chen (2011)	Australia	Interpreting devices application
Ibrahim-González (2011)	Malaysia	Interpreting devices application
Van Dyk (2010)	South Africa	Information retrieval
Orlando (2010)	Australia	Interpreting devices application
Ko (2008)	Australia	Interpreting devices application
Hansen and Shlesinger (2007)	Denmark	Interpreting devices application
Blasco Mayor et al. (2007)	Spain	Interpreting devices application
Ko (2006)	Korea	Interpreting devices application
Moser-Mercer et al. (2005)	Switzerland	Interpreting devices application
Bernstein and Isabella (2000)	America	Other interpreting technologies
Schweda-Nicholson (1985)	Canada	Interpreting devices application
Chapman (1977)	Europe	Interpreting devices application

Note: Table 3.3 follows the categories conducted by Wang and Li (2022).

3.3.2   Inherent challenges and opportunities of technology integration into interpreter education

The integration of technology into interpreter education is a multifaceted process that involves various stakeholders, including students, teachers and educational institutions. This integration presents both challenges and opportunities that vary depending on the type of technology being implemented.

Students encounter several challenges when it comes to technology integration. A significant challenge is their limited information and technology literacy, which can hinder their ability to fully utilize and benefit from interpreting technologies. As Castillo-Rodríguez et al. (2023) note, students struggle with concentration during RI teaching sessions, which can be exacerbated by technological accessibility issues and usability challenges with interpreting devices. Another notable challenge encompasses students' skepticism and reluctance toward embracing interpreting technology. Key concerns revolve around potential disengagement from the process of skill acquisition, heightened cognitive load, fears of being replaced by technology, potential compromises in the quality of interpretation and the perceived dehumanization of the interpreting process (Fantinuoli 2016; Horváth 2022). Eser et al. (2020) further point out that the use of wearable devices can cause physiological effects on students, adding to the complexity of the integration process. These challenges underscore the need for targeted support and training to address students' technological proficiency and optimize their learning experiences with interpreting technologies.

Despite these challenges, technology integration presents students with numerous opportunities. Computer application technologies, such as those studied by Yuan et al. (2023) in Malaysia, enhance student involvement by providing interactive learning platforms. Information retrieval tools, explored by Mankauskienė and Svetlana (2024) in Lithuania and the UK, improve students' language competence by granting access to vast amounts of relevant terminology and background information. Other technologies, like spoken corpora and personalized learning environments, as investigated by Xiong (2021) in China, further strengthen students' lexical knowledge and language proficiency. These examples highlight the positive impact of technology integration on enhancing students' learning outcomes in interpreter education.

Teachers, too, face challenges in adopting and effectively utilizing new technologies in interpreter education. Financial constraints, as highlighted by Chapman (1977) in his study of European interpreter education, pose a significant hurdle. The lack of suitable electronic equipment and simulated conference facilities limits teachers' ability to fully integrate technology into their teaching practices. The shift to online teaching, as evidenced by Castillo-Rodríguez et al. (2023) in Spain, has also increased teachers' workload, requiring them to adapt their teaching methods and materials. Technological barriers, such as unstable internet connections and underdeveloped infrastructure, further complicate the teaching process, as noted by Mohammed and Mustafa (2023) in their study of Iraqi interpreter education. In this case, teachers need to be proficient in utilizing interpreting technologies to facilitate virtual classes, ensuring smooth communication and effective instruction despite the distance. Additionally, teachers' relatively low technology proficiency remains a challenge that needs to be addressed (Sayaheen et al. 2023; Mohammed and Mustafa 2023).

Despite these challenges, technology integration offers teachers numerous opportunities. It enables more flexible and interactive teaching methods that enhance the learning experience for teachers (Aguirre Fernández Bravo 2020; Ko and Chen 2011; Lim 2013). Technologies also facilitate efficient data extraction and multimedia-assisted teaching, boosting interpreting teaching and training efficiency (Xiong 2021). By leveraging data extraction tools and multimedia resources, teachers can enhance their instructional strategies, provide real-time feedback and create personalized learning experiences tailored to students' needs. These opportunities not only enhance the effectiveness of interpreter education but also contribute to the professional development and growth of teachers in the field.

As for educational institutions, there is a notable gap in the incorporation of technology within interpreting education, as highlighted by Wang and Li (2022). Educational institutions encounter challenges in providing the necessary infrastructure and resources to support technology integration. Ensuring consistent grading systems for technology-enhanced projects, as discussed by Leitão (2020), can be difficult. Financial constraints limit institutions' ability to purchase and maintain advanced technology equipment, as noted by Lim (2013) in her Chinese study. Technological barriers, such as underdeveloped infrastructure, can limit the accessibility and usability of technologies, as evidenced by Mohammed and Mustafa (2023) in their Iraqi study. Challenges in curriculum design, such as adapting online teaching to large student groups and changes in professional practice, as discussed by Perramon and Ugarte (2020) in their Canadian study, also remain to be addressed. Similarly, Wan and Yuan (2022) conducted an examination of CAI training in China, revealing a scarcity of relevant courses within Bachelor of Translation and Interpreting (BTI) or Master of Translation and Interpreting (MTI) interpreter training programs.

Despite these challenges, technology integration presents institutions with opportunities. Online courses and platforms, as explored by Lee and Huh (2018) in their Korean study, expand access to interpreter education, making it more inclusive and accessible. These platforms provide a flexible learning environment that accommodates diverse student schedules and geographical locations, fostering a more diverse and globally connected interpreter education community. Technologies like virtual reality, investigated by Rudvin et al. (2022) in Italy, offer cost-effective alternatives to traditional interpreter training methods, making interpreter education more accessible to institutions within limited resources.

In addition to the challenges and opportunities discussed above, ethical considerations play a crucial role in the integration of technology into interpreter education. As technologies such as AI-assisted tools and virtual reality simulations become more prevalent, it is imperative to ensure that their use is ethical, fair and respects the privacy and dignity of all stakeholders. Data collection and utilization for AI training, for instance, must adhere to strict privacy protocols to protect students' personal information, which safeguards against potential breaches of confidentiality or misuse of sensitive data (Horváth 2022). This includes ensuring that virtual scenarios accurately reflect diverse cultural contexts and do not perpetuate harmful stereotypes or biases (Leclercq-Vandelannoitte 2017). Ensuring ethical practice in technology integration is essential for building trust and maintaining the integrity of interpreter education. Measures such as implementing clear guidelines, training programs and oversight mechanisms are advised to be taken to ensure that technology is used responsibly and ethically, fostering a safe and inclusive learning environment for all participants.

In summary, the integration of technology into interpreter education is a complex process that involves numerous challenges and opportunities for students, teachers and educational institutions. Understanding these challenges and opportunities is crucial for developing effective strategies to enhance interpreter training and address the needs of all stakeholders while overcoming technological barriers.

3.3.3 Regional differences of identified challenges and opportunities

In examining the integration of technology into interpreter education, significant regional disparities emerge, with China and Europe emerging as the leading research hubs. This reflects not only China's rapid technological and economic growth but also Europe's established history and expertise in interpreting practices. Notably, 44 out of the 67 reviewed articles originate from these two regions, highlighting their prominent roles. China, for instance, has contributed numerous studies on interpreting devices and information retrieval (Wei et al. 2022; Lim 2013), indicating its proactive engagement in technological integration. Meanwhile, Europe, with its strong tradition in interpreting, continues to innovate and adopt new technologies, focusing on diverse areas ranging from interpreting devices to other innovative tools (Castillo-Rodríguez et al. 2023; Postigo and Parrilla 2023).

A closer examination of the regional distribution reveals that Asia, led by China, produced the most research, accounting for 28 articles, with 19 specifically from China. Europe, as the birthplace of modern interpreting, provided 27 articles, with Spain standing out as a significant contributor. Oceania, represented primarily by Australia, North America and Africa also contributed, albeit to a lesser extent. This regional distribution underscores the centrality of Asia and Europe in research on integrating technology into interpreter pedagogy.

In terms of practical implementation, Europe appears to have a more mature infrastructure and resources for remote interpreter education. For instance, D'Hayer (2012) observed the implementation of virtual interpreting classes using video conference technology in Europe, primarily within the EMCI network and the interpreting services of European institutions. Furthermore, projects such as Interpreting in Virtual Reality (IVY) and EVIVA (Evaluating the education of interpreters and their clients through virtual learning activities), supported by the European Commission, demonstrate Europe's access to cutting-edge interpreting technologies and its leadership in interpreter education (Braun and Slater 2014).

In contrast, Asia, particularly central and western regions, may face limitations in developing RI technologies due to insufficient funding. Nonetheless, Asian institutions have exhibited creativity in leveraging available resources. For example, they have utilized relatively economical technologies such as speech recognition or terminology lookup systems to enhance interpreting pedagogy. Moreover, progress has been made in areas like virtual simulation lab teaching, demonstrating Asia's proactive approach to pedagogical innovation despite resource constraints. These technologies, although more cost-effective, still contribute significantly to improving interpreter training by providing tools for real-time language processing and access to essential terminology databases. This kind of relatively pragmatic approach showcases the adaptability and innovation of Asian institutions in integrating technology to overcome financial constraints and enhance interpreter education.

In assessing the impact of technological integration on interpreting pedagogy, the literature provides valuable insights. While direct comparative studies remain scarce, several pieces of research indicate that technology can enhance both teaching efficiency and the quality of student interpretation. For instance, online instruction has been shown to be as effective as traditional face-to-face teaching (e.g., Orlando and Hlavac 2020). Furthermore, the utilization of virtual reality or authentic audiovisual materials has been found to elevate students' practical readiness. The integration of automatic speech recognition technology has also been proven beneficial in improving students' interpretation accuracy and efficiency (Yu and Van Heuven 2017). These findings underscore the potential benefits of technology in interpreter education but also emphasize the need for further research to comprehensively grasp its impact.

Institutional variances in technological access and implementation are also evident. There's a clear distinction drawn between the technological access and implementation in European institutions versus those in regions like Asia. European institutions enjoy advantages stemming from funding and support from organizations like the European Commission. This is exemplified by projects such as ORCIT (Open Resources for Consecutive Interpreter Training), which evaluates the effectiveness of innovative teaching resources (Mankauskienė and Svetlana 2024). Through ORCIT and similar projects, European institutions can evaluate the effectiveness of cutting-edge teaching resources, including advanced technological tools and methodologies. This proactive approach not only enhances the quality of interpreter training, but also positions European institutions at the forefront of technology integration in interpreter education. In contrast, institutions in regions like Asia, where funding may be limited, may rely on more economical technologies to enhance teaching quality. However, efforts to leverage virtual simulation labs and other resources demonstrate these regions' proactive approach to pedagogical innovation. By creatively utilizing available technologies, Asian institutions aim to bridge the technological gap and provide students with practical and immersive learning experiences.

In summary, the integration of technology into interpreter education exhibits distinct regional and institutional variations. Europe and China lead the field, with Europe leveraging substantial funding and advanced projects, while China rapidly incorporates new technologies despite potential funding limitations. This regional diversity in technological adoption highlights the global effort to enhance interpreter pedagogy through innovative solutions, underscoring the need for further research to fully understand and harness the potential of technology in interpreter education.

3.4 Discussions

The intricate findings of this systematic review cast light on the multifaceted landscape of technology integration in interpreter education, particularly the intricate balance between the potential benefits and challenges it poses for diverse stakeholders.

The focus on student engagement and tailored learning environments underscores the evolution toward learner-centric approaches in interpreter education. Technologies, such as computer applications and information retrieval tools, hold immense potential in revolutionizing students' learning experiences by providing access to vast resources and interactive platforms that cater to their individual needs and preferences. However, the question remains as to how these technologies can be optimally designed and implemented to maximize their impact on student learning outcomes.

The challenges faced by teachers in adopting and utilizing new technologies are significant and multifaceted. Financial constraints, technological barriers and increased workload due to online teaching all contribute to the difficulty of integrating technology into teaching practices. This underscores the need for a holistic approach to teacher preparation and support, which includes not only technological literacy training but also pedagogical guidance on how to leverage technology to enhance teaching effectiveness.

The regional disparities in technological infrastructure and resources highlight the importance of a collaborative and multi-stakeholder approach to technology integration. European institutions, with their more mature infrastructure and support systems, offer valuable insights into best practices for promoting technology integration. However, the creativity and resourcefulness demonstrated by Asian institutions in leveraging limited resources also provide important lessons for other regions. A collaborative effort across regions, involving stakeholders from government, education institutions and industry, is needed to address the challenges of technological disparities and facilitate global progress in interpreter education.

The discussion of ethical considerations surrounding technological use in interpreter education points to an important yet often overlooked aspect of technology integration. As technologies such as AI-assisted tools and virtual reality simulations become more prevalent in interpreter education, it is crucial to ensure that their use is ethical, fair and respects the privacy and dignity of all stakeholders. Future research should explore innovative approaches to address these ethical challenges while harnessing the potential of technology to revolutionize interpreter education.

3.5 Conclusions and implications

This systematic review offers a comprehensive understanding of the current applications, challenges and opportunities of technology integration in interpreter education. It underscores the need for a balanced and holistic approach that addresses the diverse needs and concerns of students, teachers and educational institutions. By fostering a culture of technological experimentation and innovation, providing adequate support and resources and attending to ethical considerations, we can harness the transformative power of technology to revolutionize interpreter education, leading to more effective teaching methodologies and ultimately improved interpreter training quality. The future of interpreter education lies in sustained research, collaboration and innovation in technology integration. This requires a concerted effort from all stakeholders to ensure that technology is leveraged optimally to benefit the interpreter education landscape.

References

Aguirre Fernández Bravo, Elena. 2020. "The Impact of ICT on Interpreting Students' Self-Perceived Learning." In *The Role of Technology in Conference Interpreter Training*, edited by Rodríguez Melchor María Dolores, Ildikó Horvath, and Kate Ferguson, 203–20. New York: Peter Lang. https://doi.org/10.3726/b13466.

Ahrens, Barbara, Morven Beaton-Thome, and Anja Rütten. 2021. "The Pivot to Remote Online Teaching on the MA in Conference Interpreting in Cologne: Lessons learned from an Unexpected Experience." *The Journal of Specialised Translation* 36: 251–84.

Alcaide-Martínez, Marta, and Lidia Taillefer. 2022. "Gamification for English Language Teaching: A Case Study in Translation and Interpreting." *Lebende Sprachen* 67(2): 283–310. https://doi.org/10.1515/les-2022-1015.

Amos, Rhona M., Kilian G. Seeber, and Martin J. Pickering. 2023. "Student Interpreters Predict Meaning While Simultaneously Interpreting – Even Before Training." *Interpreting* 25(2): 211–38. https://doi.org/10.1075/intp.00093.amo.

Arzık-Erzurumlu, Ozum. 2020. "Employing Podcasts as a Learning Tool in Interpreter Training: A Case Study." In *The Role of Technology in Conference Interpreter Training*, edited by Rodríguez Melchor María Dolores, Ildikó Horvath, and Kate Ferguson, 179–203. New York: Peter Lang. https://doi.org/10.3726/b13466.

Bale, Richard. 2013. "Undergraduate Consecutive Interpreting and Lexical Knowledge: The Role of Spoken Corpora." *The Interpreter and Translator Trainer* 7(1): 27–50.

Bernstein, Jared and Isabella Barbier. 2000. "Design and Development Parameters for a Rapid Automatic Screening Test for Prospective Simultaneous Interpreters." *Interpreting* 5(2): 221–38.

Blasco, Mayor, María Jesús, and María Amparo Jiménez Ivars. 2007. "E-Learning for Interpreting." *Babel* 53(4): 292–302.

Braun, Sabine. 2019. "Technology and Interpreting." In *The Routledge Handbook of Translation and Technology*, edited by Minako O'Hagan, 271–89. London: Routledge.

Braun, Sabine, Elena Davitti, and Catherine Slater. 2020. "'It's Like Being in Bubbles': Affordances and Challenges of Virtual Learning Environments for Collaborative Learning in Interpreter Education." *The Interpreter and Translator Trainer* 14(3): 1–20. https://doi.org/10.1080/1750399X.2020.1800362.

Braun, Sabine, and Catherine Slater. 2014. "Populating a 3D Virtual Learning Environment for Interpreting Students with Bilingual Dialogues to Support Situated Learning in an Institutional Context." *The Interpreter and Translator Trainer* 8(3): 469–85. https://doi.org/10.1080/1750399X.2014.971484.

Castillo-Rodríguez, Cristina, Cristina Toledo-Báez, and Miriam Seghiri. 2023. "Teaching Interpreting in Times of Covid: Perspectives, Experience and Satisfaction." *Revista de Lingüística y Lenguas Aplicadas* 18: 19–33. https://doi.org/10.4995/rlyla.2023.18747.

Chan, Venus. 2023a. "Investigating the Impact of a Virtual Reality Mobile Application on Learners' Interpreting Competence." *Journal of Computer Assisted Learning* 39: 1242–58. https://doi.org/10.1111/jcal.12796.

Chan, Venus. 2023b. "Research on Computer-Assisted Interpreter Training: A Review of Studies from 2013 to 2023." *SN Computer Science* 4: 648. https://doi.org/10.1007/s42979-023-02072-w.

Chapman, Craig. 1977. "Applications of the Language Laboratory to Training in Simultaneous Interpretation." *Meta: Translators' Journal* 22: 264–69.

Chen, Sijia. 2020. "The process of Note-taking in Consecutive Interpreting: A Digital Pen Recording Approach." *Interpreting* 22(1): 117–39.

Costa, Hernani, Gloria Corpas Pastor, and Isabel Durán-Muñoz. 2014. "Technology-Assisted Interpreting." *Multilingual* 143(25): 27–32.

Crezee, Ineke Hendrika, Jo Anna Burn, and Nidar Gailani. 2015. "Authentic Audiovisual Resources to Actualise Legal Interpreting Education." *MonTI – Monografías de Traducción e Interpretación* 7: 271–93. https://doi.org/10.6035/MONTI.2015.7.10.

D'Hayer, Danielle. 2012. "Public Service Interpreting and Translation: Moving Towards a (Virtual) Community of Practice." *Meta* 57(1): 235–47. https://doi.org/10.7202/1012751ar.

Eraslan, Şeyda, Mehmet Şahin, Gazihan Alankuş, Özge Altıntaş, and Damla Kaleş. 2020. "Virtual Worlds as a Contribution to Content and Variety in Interpreter Training: The Case of Turkey." In *The Role of Technology in Conference Interpreter Training*, edited by Rodríguez Melchor María Dolores, Ildikó Horvath, and Kate Ferguson, 101–28. New York: Peter Lang. https://doi.org/10.3726/b13466.

Eser, Oktay, Miranda Lai, and Fatih Saltan. 2020. "The Affordances and Challenges of Wearable Technologies for Training Public Service Interpreters." *Interpreting* 22: 288–308. https://doi.org/10.1075/intp.00044.ese.

Fantinuoli, Claudio. 2016. "InterpretBank: Redefining Computer-assisted Interpreting Tools." In *Proceedings of the 38th Conference Translating and the Computer*, edited by João Esteves-Ferreira, Juliet Margaret Macan, Ruslan Mitkov, and Olaf-Michael Stefanov, 42–52. London: AsLing.

Gerber, Leah, Jim Hlavac, Irwyn Shepherd, Paul McIntosh, Alex Avella Archila, and Hyein Cho. 2021. "Stepping into the Future: Virtual Reality Training for Community Interpreters Working in the Area of Family Violence." *The Journal of Specialized Translation* 36: 252–75.

Goldsmith, Joshua. 2018. "Tablet Interpreting: Consecutive Interpreting 2.0." *Translation and Interpreting Studies* 13(3): 342–65. https://doi.org/10.1075/tis.00020.gol.

Han, Chao. 2021. "Analytic Rubric Scoring Versus Comparative Judgment: A Comparison of Two Approaches to Assessing Spoken-language Interpreting." *Meta* 66(2): 337–61.

Han, Lili, Yuying Wang, and Yumeng Li. 2022. "Student Perceptions of Online Interpreting Teaching and Learning via the Zoom Platform." *Teaching English as a Second or Foreign Language–TESL-EJ* 26 (1): 1–19. https://doi.org/10.55593/ej.26101int.

Hansen, Inge Gorm, and Miriam Shlesinger 2007. "The Silver Lining. Technology and Self-Study in the Interpreting Classroom." *Interpreting* 9(1): 95–118. https://doi.org/10.1075/intp.9.1.06gor.

Horváth, Ildikó. 2022. "AI in Interpreting: Ethical Considerations." *Across Languages and Cultures* 23(1): 1–13. https://doi.org/10.1556/084.2022.00108.

Hu, Zhiwen, and Ian McGrath. 2012. "Integrating ICT into College English: An Implementation Study of a National Reform." *Education and Information Technologies* 17: 147–65. https://doi.org/10.1007/s10639-011-9153-0.

Huang, Yuwei, Weinan Shi, and Jinglin Wen. 2023. "Technology Challenges and Aids: The Sustainable Development of Professional Interpreters in Listening Comprehension Effectiveness and Interpreting Performance." *Sustainability* 15(8): 6828. https://doi.org/10.3390/su15086828.

Hunt-Gómez, Coral, and Gómez Moreno Paz. 2015. "Reality-Based Court Interpreting Didactic Material Using New Technologies." *The Interpreter and Translator Trainer* 9(2): 188–204. https://doi.org/10.1080/1750399X.2015.1051770.

Huỳnh, Thi Lan, and Uyen Nu Thuy Nguyen. 2019. "Students' Perceptions and Design Considerations of Flipped Interpreting Classroom." *Theory and Practice in Language Studies* 9(9): 1100–10. https://doi.org/10.17507/tpls.0909.05.

Ibrahim-González, Noraini. 2011. "E-Learning in Interpreting Didactics: Students' Attitudes and Learning Patterns, and Instructor's Challenges." *The Journal of Specialized Translation* 16: 224–41.

Jekat, Susanne. 2015. "Machine interpreting". In *Routledge Encyclopedia of Interpreting Studies*, edited by Franz Pöchhacker, 239–41. London: Routledge.

Jiang, Kai, and Xi Lu. 2021. "The Influence of Speech Translation Technology on Interpreter's Career Prospects in the Era of Artificial Intelligence." *Journal of Physics: Conference Series* 1802(4): 042074. https://doi.org/10.1088/1742-6596/1802/4/042074.

Ko, Leong. 2006. "Teaching Interpreting by Distance Mode: Possibilities and Constraints." *Interpreting* 8(1): 67–96. https://doi.org/10.1075/INTP.8.1.05KO.

Ko, Leong. 2008. "Teaching Interpreting by Distance Mode: An Empirical Study." *Meta* 53(4): 814–40. https://doi.org/10.7202/019649ar.

Ko, Leong, and Nian-Shing Chen 2011. "Online-Interpreting in Synchronous Cyber Classrooms." *Babel* 57(2): 123–43. https://doi.org/10.1075/BABEL.57.2.01KO.

Leclercq-Vandelannoitte, Aurelie. 2017. "An Ethical Perspective on Emerging Forms of Ubiquitous IT-Based Control." *Journal of Busineess Ethics* 142(1): 139–54. https://doi.org/10.1007/s10551-015-2708-z.

Lee, Jieun, and Jiun Huh 2018. "Why Not Go Online?: A Case Study of Blended Mode Business Interpreting and Translation Certificate Program." *The Interpreter and Translator Trainer* 12(4): 444–66. https://doi.org/10.1080/1750399X.2018.1540227.

Lei, Ming, and Yong Zhu. 2013. "On the Multi-Media Assisted Interpreting Teaching Strategies." In *Informatics and Management Science II*, edited by Wenjiang Du, 601–08. London: Springer. https://doi.org/10.1007/978-1-4471-4811-1_77.

Leitão, Fernando. 2020. "The Speech Repository: Challenges and New Projects." In *The Role of Technology in Conference Interpreter Training*, edited by Rodríguez Melchor María Dolores, Ildikó Horvath, and Kate Ferguson, 43–58. New York: Peter Lang. https://doi.org/10.3726/b13466.

Li, Xi, and Qing Jiang. 2021 "Application of Virtual Reality Technology in English Interpretation Teaching." In *2nd International Conference on Information Science and Education (ICISE-IE)*, 124–27. Chongqing, China. https://doi.org/10.1109/ICISE-IE53922.2021.00035.

Li, Saihong, Wang Yifang, and Rasmussen Yubo Zhou. 2022. "Studying Interpreters' Stress in Crisis Communication: Evidence from Multimodal Technology of Eye-Tracking, Heart Rate and Galvanic Skin Response." *The Translator* 28(4): 468–88. https://doi.org/10.1080/13556509.2022.2159782.

Lim, Lily. 2013. "Examining Students' Perceptions of Computer-Assisted Interpreter Training." *The Interpreter and Translator Trainer* 7(1): 71–89. https://doi.org/10.1080/13556509.2013.798844.

Lim, Lily. 2020. "Interpreting Training in China: Past, Present, and Future." In *Key Issues in Translation Studies in China*, edited by Lily Lim and Defeng Li, 143–59. Singapore: Springer. https://doi.org/10.1007/978-981-15-5865-8_7.

Ludvigsen, Mette S., Elisabeth O. C. Hall, Gabriele Meyer, Liv Fegran, Hanne Aagaard, and Lisbeth Uhrenfeldt. 2016. "Using Sandelowski and Barroso's Meta-Synthesis Method in Advancing Qualitative Evidence." *Qualitative Health Research* 26(3): 320–29. https://doi.org/10.1177/1049732315576493.

Lu, Xiaolei, and Chao Han. 2022. "Automatic Assessment of Spoken-Language Interpreting Based on Machine-Translation Evaluation Metrics: A Multi-Scenario Exploratory Study." *Interpreting* 25(1): 109–43.

Mankauskienė, Dalia, and Carsten Svetlana. 2024. "Evaluating the Effectiveness of ORCIT's Note-Taking Resource." *Onomázein Revista de lingüística filología y traducción*: 39–58. https://doi.org/10.7764/onomazein.ne13.03.

Manko, Volodymyr M., and Liudmyla I. Khitsenko. 2019. "The Use of Multimedia Linguistic Laboratories in Foreign Language Learning (the Case Study of «Translation» Specialty)." *Information Technologies and Learning Tools* 74(6): 201–11. https://doi.org/10.33407/itlt.v74i6.2592.

Metruk, Rastislav. 2022. "Smartphone English Language Learning Challenges: A Systematic Literature Review." *Sage Open* 12(1): 1–15. https://doi.org/10.1177/21582440221079627.

Mohammed, Hala, and Balsam Mustafa. 2023. "Forced to Go Online: A Case Study of Learning Consecutive and Simultaneous Interpreting under Covid-19 in Iraq." *Translation and Interpreting* 15(1): 176–99.

Moher, David, Alessandro Liberati, Jennifer Tetzlaff, Douglas G. Altman, and The PRISMA Group. 2009. "Preferred Reporting Items for Systematic Reviews and Meta-Analyses: The PRISMA Statement." *PLoS Medicine* 6(7): e1000097. https://doi.org/10.1371/journal.pmed.1000097.

Moser-Mercer, Barbara, Barbara Class, and Kilian G. Seeber. 2005. "Leveraging Virtual Learning Environments for Training Interpreter Trainers." *Meta: Translators' Journal* 50(4): 1–25.

Nai, Ruihua and Hanita Hassan. 2023. "Application of Multimodal Speech Recognition Based on Deep Neural Networks in Interpretation Teaching," In *Proc. SPIE 12923, Third International Conference on Artificial Intelligence, Virtual Reality, and Visualization (AIVRV 2023)*, 129230N (8 November 2023). https://doi.org/10.1117/12.3011751

Nai, Ruihua and Hanita Hassan. 2022. "Multi-modal Simultaneous Interpreting Teaching: Based on Situated Learning in Virtual Reality." In *Innovative Computing*, edited by Yiyang Pei, Jia Wei Chang, and Jason C. Hung, 494–501. IC 2022. Lecture Notes in Electrical Engineering, vol. 935. Singapore: Springer. https://doi.org/10.1007/978-981-19-4132-0_61

Orlando, Marc. 2010. "Digital Pen Technology and Consecutive Interpreting: Another Dimension in Notetaking Training and Assessment." *Interpreters Newsletter* 15: 71–86.

Orlando, Marc. 2014. "A Study on The Amenability of Digital Pen Technology in A Hybrid Mode Of Interpreting: Consec-Simul with Notes." *Translation and Interpreting*, 6(2): 39–54.

Orlando, Marc. 2015. "Digital Pen Technology and Interpreter Training, Practice and Research: Status and Trends." In *Interpreter Education in the Digital Age*, edited by Suzanne Ehrlich and Jemina Napier, 125–52. Washington: Gallaudet University Press.

Orlando, Marc, and Jim Hlavac 2020. "Simultaneous-Consecutive in Interpreter Training and Interpreting Practice: Use and Perceptions of a Hybrid Mode." *The Interpreters' Newsletter* 25: 1–17.

Peng, Gracie. 2022. "Exploring Automatic Speech Recognition Technology for Undergraduate Sight Translation Training." *Compilation and Translation Review* 15(2): 199–242.

Perramon, María, and Xus Ugarte Ballester. 2020. "Teaching Interpreting Online for the Translation and Interpreting Degree at the University of Vic: A Nonstop Challenge Since 2001." *Translation and Translanguaging in Multilingual Contexts* 6(2): 172–82. https://doi.org/10.1075/ttmc.00052.per.

Petite, Christelle. 2005. "Evidence of Repair Mechanisms in Simultaneous Interpreting: A Corpus-based Analysis." *Interpreting* 7(1): 27–49.

Petticrew, Mark, and Helen Roberts. 2006. *Systematic Reviews in the Social Sciences: A Practical Guide*. Oxford: Blackwell.

Postigo Pinazo, Encarnación, and Laura Parrilla Gómez. 2023. "Analysis of Audio Transcription Tools with Real Corpora: Are They a Valid Tool for Interpreter Training?" In *New Trends in Healthcare Interpreting Studies,* edited by Raquel Lázaro Gutiérrez and Álvaro Aranda Cristina, 173–87. Singapore: Springer. https://doi.org/10.1007/978-981-99-2961-0_9.

Prandi, Bianca. 2020. "The Use of CAI Tools in Interpreter Training: Where Are We Now and Where Do We Go from Here?" In *TRAlinea Special Issue: Technology in Interpreter Education and Practice*. Accessed January 24, 2023. https://www.intralinea.org/specials/article/2512.

Qassem, Mutahar, and Buthainah M. Al Thowaini. 2024. "Effectiveness of an Online Training Course for Trainee Translators: Analysis of Keylogging Data." *Education and Information Technologies* 29: 15711–35. https://doi.org/10.1007/s10639-024-12484-7.

Rodríguez Melchor, María Dolores. 2020. "Meeting the Challenge of Adapting Interpreter Training and Assessment to Blended Learning Environments." In *The Role of Technology in Conference Interpreter Training*, edited by Rodríguez Melchor María Dolores, Ildikó Horvath, and Kate Ferguson, 59–76. New York: Peter Lang. https://doi.org/10.3726/b13466.

Rodríguez Melchor, María Dolores, Manuela Motta, Elena Aguirre Fernández Bravo, Olga Egorova, Kate Ferguson, and Tamara Mikolič Južnič. 2020. "Expertise and Resources for Interpreter Training Online: A Student Survey on Pedagogical and Technical Dimensions of Virtual Learning Environments." *Babel* 66(6): 950–72.

Rudvin, Mette, Edoardo Di Gennaro, and Roberta Teresa Di Rosa. 2022 "Training Language Mediators and Interpreters through Embodied Cognition, Immersive Learning and Virtual Reality: Didactic, Organizational and Cost Benefits." *Italian Journal of Sociology of Education* 14(3): 131–52. https://doi.org/10.14658/pupj-ijse-2022-3-6.

Ruiz Mezcua, Aurora. 2019. "El Triple Reto de la Interpretación a Distancia: Tecnológico, Profesional y Didáctico." *MonTI: Monografías de Traducción e Interpretación* 11: 43–262.

Şahin, Mehmet. 2013. "Virtual Worlds in Interpreter Training." *The Interpreter and Translator Trainer* 7(1): 91–106. https://doi.org/10.1080/13556509.2013.10798845.

Saldaña, Johnny. 2016. *The Coding Manual for Qualitative Researchers.* Los Angeles, CA: Sage.

Sayaheen, Bilal, Mohannad Sayaheen, and Ibrahim Darwish. 2023. "Teaching Translation Technology in the Age of COVID-19: An Analysis of Data Gathered in Jordanian Universities." *New Voices in Translation Studies* 28(2): 52–77. https://doi.org/10.14456/nvts.2023.20.

Schweda-Nicholson, Nancy. 1985. "Consecutive Interpretation Training: Videotapes in the classroom." *Meta* 30(2): 148–54. https://doi.org/10.7202/003731ar.

Seeber, Kilian G., and Carmen Delgado Luchner. 2020. "Simulating Simultaneous Interpreting with Text: From Training Model to Training Module." In *The Role of Technology in Conference Interpreter Training*, edited by Rodríguez Melchor María Dolores, Ildikó Horvath, and Kate Ferguson, 129–51. New York: Peter Lang. https://doi.org/10.3726/b13466.

Seresi, Márta, and Ildikó Horváth. 2020. "Virtual Classes: Students' and Trainers' Perspectives." In *The Role of Technology in Conference Interpreter Training*, edited by Rodríguez Melchor María Dolores, Ildikó Horvath, and Kate Ferguson, 155–78. New York: Peter Lang. https://doi.org/10.3726/b13466.

Shaffril, Hayrol Azril Mohamed, Nobaya Ahmad, Samsul Farid Samsuddin, Asnarulkhadi Abu Samah, and Mas Ernawati Hamdan. 2020. "Systematic Literature Review on Adaptation towards Climate Change Impacts among Indigenous People in the Asia Pacific Regions." *Journal of Cleaner Production* 258: 1–14. https://doi.org/10.1016/j.jclepro.2020.120595.

Slettebakk Berge, Sigrid. 2018. "How Sign Language Interpreters Use Multimodal Actions to Coordinate Turn-Taking in Group Work between Deaf and Hearing Upper Secondary School Students." *Interpreting* 20(1): 96–125.

Su, Wenchao, and Defeng Li. 2021. "Exploring the Effect of Interpreting Training: Eye-Tracking English-Chinese Sight Interpreting." *Lingua* 256: 103094. https://doi.org/10.1016/j.lingua.2021.103094.

Svetlana, Carsten, Nijolé Maskaliūnienė, and Matthew Perret. 2020. "The Collaborative Multilingual Multimedia Project ORCIT (Online Resources in Conference Interpreter Training): Sharing Pedagogical Good Practice and Enhancing Learner Experience." In *The Role of Technology in Conference Interpreter Training*, edited by Rodríguez Melchor María Dolores, Ildikó Horvath, and Kate Ferguson, 77–100. New York: Peter Lang. https://doi.org/10.3726/b13466.

Tian, Xianzhi. 2014. "Exploitation and Application of Information System in Interpretation Training." *Advanced Materials Research* 1014: 434–37. https://doi.org/10.4028/www.scientific.net/amr.1014.434.

Tian, Sha, and Wenjiao Yang 2023. "Modeling the Use Behavior of Interpreting Technology for Student Interpreters: An Extension of UTAUT Model." *Education and Information Technologies* 29(9): 1–30. https://doi.org/10.1007/s10639-023-12225-2.

Valero Garcés, Carmen. 2020. "Creating a Virtual Learning Environment in Public Service Interpreting and Translation. The Massive Open Online Course Get Your Start in PSIT." *The Interpreters' Newsletter* 25: 37–48. https://doi.org/10.13137/2421-714X/31236.

Van Dyk, Jeanne. 2010. "Multilingual News Websites as a Resource for Interpreter Training." *Southern African Linguistics and Applied Language Studies* 28(3): 291–97. https://doi.org/10.2989/16073614.2010.545031.

Wan, Hongyu, and Xiaoshu Yuan. 2022. "Perceptions of Computer-Assisted Interpreting Tools in Interpreter Education in Chinese Mainland: Preliminary Findings of a Survey." *International Journal of Chinese and English Translation and Interpreting* 1: 1–28. https://doi.org/10.56395/ijceti.v1i1.8.

Wang, Yuxi, Liping Chen, and Jiayin Han. 2024. "Exploring Factors Influencing Students' Willingness to Use Translation Technology." *Education and Information Technologies* 29: 17097–118. https://doi.org/10.1007/s10639-024-12511-7.

Wang, Huashu 王华树, and Yang Chengshu 杨承淑. 2019. "Rengongzhineng shidai de kouyi jishu fazhan: gainian, yingxiang yu qushi" 人工智能时代的口译技术发展：概念、影响与趋势 [Interpreting Technologies in the Era of Artificial Intelligence: Concepts, Influences and Trends.] *Zhongguo fanyi* 中国翻译 [*The Chinese Translators Journal*] 40(6): 69–79.

Wang, Huashu, and Zhi Li. 2022. "Constructing a Competence Framework for Interpreting Technologies, and Related Educational Insights: an Empirical Study." *The Interpreter and Translator Trainer* 16(3): 367–90. https://doi.org/10.1080/1750399X.2022.2101850.

Wang, Yuying, Yunan Tian, Yunxiao Jiang, and Zhonggen Yu. 2023. "The Acceptance of Tablet for Note-Taking in Consecutive Interpreting in a Classroom Context: The Students' Perspectives." *Forum for Linguistic Studies* 5(2): 1862. https://doi.org/10.59400/fls.v5i2.1862.

Wei, Wei, Yi Yu, and Ge Gao. 2022. "Investigating Learners' Changing Expectations on Learning Experience in a MOOC of Professional Translation and Interpreter Training." *Sage Open* 12(4). https://doi.org/10.1177/21582440221134577.

Xiao, Yu, and Maria Watson. 2019. "Guidance on Conducting a Systematic Literature Review." *Journal of Planning Education and Research* 39(1): 93–112. https://doi.org/10.1177/0739456X17723971.

Xiong, Huiqin. 2021. "Application of Computer Big Data Technology in the Teaching of Interpretation and Translation." In *2021 4th International Conference on Information Systems and Computer Aided Education (ICISCAE'21),* 726–30. https://doi.org/10.1145/3482632.3483004.

Xu, Ran. 2018. "Corpus-Based Terminological Preparation for Simultaneous Interpreting." *Interpreting* 20(1): 29–58. https://doi.org/10.1075/intp.00002.xu.

Yang, Ping. 2022. "Novel Interpretation Teaching and its Implementation Paths Under Informatization and Big Data." In *International Conference on Computation, Big-Data and Engineering (ICCBE),* 158–61. Yunlin, Taiwan. https://doi.org/10.1109/ICCBE56101.2022.9888221.

Yang, Yanxia. 2024. "Understanding Machine Translation Fit for Language Learning: The Mediating Effect of Machine Translation Literacy." *Education and Information Technologies* 29: 20163–180. https://doi.org/10.1007/s10639-024-12650-x.

Ye, Hui. 2021. "Digital Technology-Based Pedagogy for Interpreting." In *Application of Intelligent Systems in Multi-modal Information Analytics,* edited by Vijayan Sugumaran, Zheng Xu, and Huiyu Zhou, 521–27. https://doi.org/10.1007/978-3-030-51431-0_76

Yu, Wenting, and Vincent J. Van Heuven. 2017. "Predicting Judged Fluency of Consecutive Interpreting from Acoustic Measures: Potential for Automatic Assessment and Pedagogic Implications." *Interpreting* 19(1): 47–68.

Yuan, Yihuan, Jamalludin Harun, and Zhiru Wang. 2023. "The Effects of Mobile-Assisted Collaborative Language Learning on EFL Students' Interpreting Competence and Motivation." *International Journal of Computer-Assisted Language Learning and Teaching* 13(1): 1–19. https://doi.org/10.4018/IJCALLT.332404.

Zhao, Yihui 赵毅慧. 2017. "Kouyi jishu de huisu yu qianzhan: gongjuhua jiaohuhua ji zhinenghua de yanbian" 口译技术的回溯与前瞻：工具化′ 交互化及智能化的演变 [A Review of the Development of Interpreting Technologies: A Shift from Instrumentality to Interactivity, and to Intelligence]. *Waiwen yanjiu* 外文研究 [*Foreign Studies*] 5(4): 65–71, 105.

Zhong, Yong, Jiacheng Xie, and Ting Zhang. 2021. "Crowd Creation and Learning of Multimedia Content: An Action Research Project to Create Curriculum 2.0 Translation and Interpreting Courses." *The Interpreter and Translator Trainer* 15(1): 85–101. https://doi.org/10.1080/1750399X.2021.1880695.

4

APPLICATION OF SKETCH ENGINE AND ANTCONC IN COMPARING FEATURES OF LEXICAL BUNDLES IN INTERPRETED AND NATIVE CHINESE

Dan Feng Huang and Fei Deng

4.1 Introduction

The aim of this chapter is to empirically investigate the extent to which the corpus technologies (CTs) of Sketch Engine (SE) (Kilgarriff et al. 2014) and AntConc (Anthony 2005, 2019) can use features of lexical bundles (LBs) to distinguish interpreted and native Chinese (Mandarin). Since Baker (1993) proposed the concept of translation universals and a corpus linguistic approach to study the concept, there have been extensive corpus-based translation studies (CBTS) to identify the distinctive features in translated texts (Biel et al. 2019; Fu and Wang 2022; Liu and Afzaal 2021; Liu et al. 2023; Su et al. 2023; Xiao 2011). Relevant studies can also be found in edited books (Cheung et al. 2023; Mauranen and Kujamäki 2004) and a review by Granger and Lefer (2022). These studies have used CTs such as WordSmith Tools (Scott 2001), AntConc or more advanced automatic annotation and extraction techniques. They have provided empirical evidence to contribute to the understanding of translated language as a special norm or "third code" (Frawley 1984, 173).

While it is acknowledged that CBTS have facilitated a nuanced understanding of a wide range of translation phenomena (Cheung et al. 2023), an interesting issue that has not been explored is whether and how different CTs would achieve the same research purpose when applied to investigate the same linguistic phenomenon. CTs have gone through four generations of development (Hardie 2012; Li and Guo 2016). The first generation relies on mainframe computers, which are large, powerful and centrally controlled computers owned by organizations and institutions. An example is the CLOC (Concordance Locator) developed by Reed in 1978 (Reed 1978) and used at the University of Birmingham. The second generation uses PC software such as the Kaye Concordance (Kaye 1990). The third

DOI: 10.4324/9781003597711-5

generation uses desktop PCs; they provide a wider range of functions for linguistic studies than the second generation and do not require programming skills, as exemplified by WordSmith, MonoConc (Barlow 2000) and AntConc. The most recent generation as of the completion of this chapter is web-based. The web interface is used to communicate the researcher's needs to the Corpus Query Processor, which uses computational algorithms and global and large-scale corpora to provide data to satisfy the researcher's needs. The fourth generation is called CQPweb and some well-known CTs are Wmatrix (Rayson 2008) and SE. Like the third generation, the fourth generation does not require programming skills.

Hardie (2012) proposed that the fourth generation is expected to maintain a balance between power and usability. Power refers to its ability to process data effectively, while usability refers to its user-friendliness. SE strikes for such a balance. It provides access to worldwide corpora for multilingualism and diverse genres and registers. It offers multiple functions, eliminating the need for users to install software on their PCs and enabling them to upload and create their corpora. Additionally, it provides automatic annotation and statistical measures for linguistic analysis. Most importantly, annotation and calculation are performed through backend computations, releasing linguistics from the pressure of programming. Despite these functions, SE has yet to be fully applied in CBTS. Granger and Lefer (2022, 27) have pointed out that "surprisingly, only two (corpus-based translation and interpreting) studies make use of Sketch Engine".

In this chapter, an attempt is made to apply SE to the analysis of LBs in interpreted Chinese and native Chinese. Chinese, as a "large language" (Kilgarriff et al., 2014, 17) with many speakers, is addressed by SE. SE has the function of n-grams for LBs. LBs are uninterrupted word clusters that are automatically extracted in a statistically driven approach (Biber et al. 2004; Stubbs 2007). LBs can be of different lengths and can be referred to as n-grams. There has been a trend to distinguish translated from non-translated texts by LBs in CBTS (Biel 2018; Biel et al. 2019; Huang et al. 2023; Lee 2013; Liu 2023; Wu 2021). Some studies have been conducted in the Chinese context (Liu et al. 2022; Xiao 2011) and have focused on written translation rather than interpreting. It is believed that LBs are prevailing in interpreting (Aston 2018; Henriksen 2007; Jones 1998). In interpreting, there have been studies examining English LBs (Huang et al. 2023; Liu 2023; Wu 2021), and a few have examined Chinese LBs (Wu et al. 2021). As noted by Xiao et al. (2010, 184), to discover translation universals, "the language pairs involved must not be limited to English and closely related languages". Thus, the inclusion of a typologically distant language such as Chinese is essential. Therefore, in this chapter, "interpretese" (Shlesinger 2008, 250) is expected to be identified when SE is used to analyze LBs in the Chinese context.

This chapter is methodological. On the one hand, it contributes to the application of SE in CBTS. On the other hand, when SE is compared with the other CTs for the same research purpose, the results provide insight into the question proposed at the beginning of the study, which makes this chapter methodologically oriented: whether and how different CTs would achieve the same research purpose when applied to investigate the same linguistic phenomenon. The latest version of AntConc 4.1.1 is chosen for comparison with SE because they have a similar function of studying LBs. A detailed comparison is presented in the following sections. The comparison may highlight the role that CTs play in CBTS.

4.2 Research background

This section provides relevant details about analyzing LBs in SE and AntConc for background information and continues to synthesize CBTS on LBs to demonstrate how and why Chinese LBs are analyzed in this chapter. Lastly, methodologically oriented studies on applications of SE and AntConc are summarized to highlight the significance of using the two CTs simultaneously to study LBs in native and interpreted Chinese.

4.2.1 A comparison between SE and AntConc

AntConc is a freely available software tool initially designed for English academic writing and later widely adopted for corpus-based linguistic research and studies (Afanasev 2020; Götz 2021; He 2015; Li and Guo 2016). Developed and released by Laurence Anthony in 2002, it has seen more than 30 historical versions since then. Detailed introductions to the tool's functions and guidance on using the tool can be found on Laurence Anthony's website.[1] Inspiration for applying AntConc to corpus linguistic research and classroom teaching can be obtained from Anthony's publications (Anthony 2005, 2006, 2019). SE, launched by Kilgarriff and Rychlý in 2004, is a lexicography tool that offers word sketches encompassing collocational and grammatical behaviors of words (Kilgarriff et al. 2014). It helps researchers and learners explore lexical patterns of words (Huang et al. 2005; Kilgarriff et al. 2015; Pearce 2008; Wang and Huang 2013). The official SE website[2] provides a tutorial and user's guide to support effective utilization. Users can also access a one-month free trial to explore its functions. Thomas (2017) dedicates a book to demonstrate how to use SE to discover language patterns in English. All these sources serve as references for comparing the two CTs in the following part.

AntConc and SE have identical functions: First, they have Concordance to present contexts of words and sources of texts. In AntConc, KWIC (KeyWord In Context) is designed for this function, and in SE, it is Concordance; Second, the two CTs can generate a wordlist and a keyword list of a text and provide visual representations for the keywords to show aboutness of the text; Third, they can explore collocates and co-occurring behaviors with an individual word as a node. The most relevant function for this chapter is the generation of LBs based on frequency and dispersion in texts. They both name the function *n*-gram ("*n*-grams" in SE and "*n*-gram" in AntConc). Therefore, in this chapter, LBs are considered the same as *n*-grams. The following two figures present the same function in two CTs.

KWIC   Plot   File   Cluster   N-Gram   Collocate   Word   Keyword   Wordcloud

N-Gram Types 0 **N-Gram Tokens** 0 **Page Size** 100 hits ⌄ ⟳ 0 to 0 of 0 hits   ⟳

Type Rank Freq Range

Search Query ☑ Words ☐ Case ☐ Regex **N-Gram Size** 4 ⇕ **Open Slots** 0 ⇕ **Min. Freq** 1 ⇕ **Min. Range** 3 ⇕

                                              ⌄   Start   ☐ Adv Search

Sort by Frequency ⌄ ☐ Invert Order

FIGURE 4.1   *n*-Gram in AntConc.

As can be seen in Figure 4.1, types and tokens are presented. The size, minimum frequency and range of texts can be set in advance. Uppercase and lowercase can be distinguished. Regex (regular expression) can define complex search patterns. After the generation of LBs, statistics, including absolute frequency, range, normalized frequency and normalized range are provided for further analysis. Normalized frequency expresses the frequency of a lexical per million words, thus allowing for comparisons of frequencies of LBs across corpora of different sizes. Normalized range is calculated by dividing the raw range by the expected range if an LB is evenly distributed throughout the corpus (the total number of texts in a corpus). The result is a value between 0 and 1. The closer a value is to 1, the more likely an LB is distributed in different texts in a corpus.

Figure 4.2 shows that there are many options in the advanced search of SE. Researchers can determine the size and the minimum frequency of LBs. In addition, they can choose word *n*-grams or part of speech (POS) *n*-grams in "Attribute" and determine the appearances of LBs through "Include nonword", "Exclude these words" and additional criteria. There are two unique functions: nest *n*-grams and key-*n*-grams. Nest *n*-grams refer to the subordinate *n*-grams for their sub-groups. For example, "he did not" is the nest *n*-gram for "did not" and "he did". Key *n*-grams refer to the *n*-grams that only exist in a specific corpus with a larger corpus as the reference corpus. Instead of defining one specific length at one time of retrieval, as AntConc does, SE allows the extraction of LBs of two- to six-word length at one time. After the generation of LBs, statistics for further analysis include (raw) frequency, relative frequency, document frequency (DOCF), relative DOCF, average

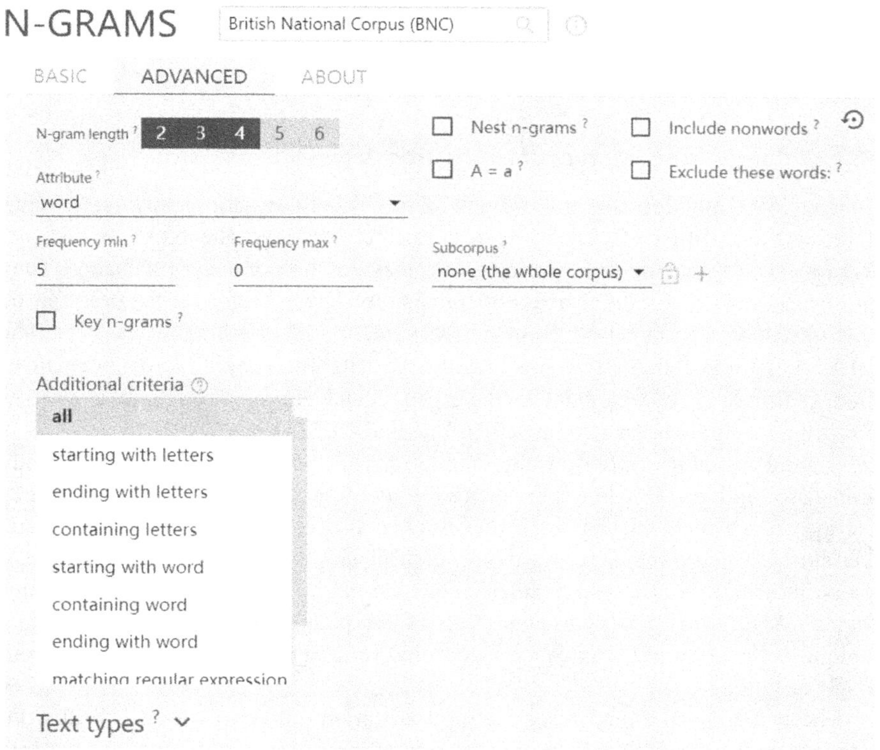

N-GRAMS

British National Corpus (BNC)

BASIC   ADVANCED   ABOUT

N-gram length ? | 2 | 3 | 4 | 5 | 6   ☐ Nest n-grams ?   ☐ Include nonwords ?

Attribute ?
word ▼   ☐ A = a ?   ☐ Exclude these words: ?

Frequency min ?   Frequency max ?   Subcorpus ?
5   0   none (the whole corpus) ▼ 🔒 +

☐ Key n-grams ?

Additional criteria ⓘ

all

starting with letters

ending with letters

containing letters

starting with word

containing word

ending with word

matching regular expression

Text types ? ⌄

FIGURE 4.2 Advanced searching *n*-grams in SE.

reduced frequency (ARF) and average logarithmic distance frequency (ALDF). According to information on the official website of SE, relative frequency refers to frequency per millions and relative DOCF means the percentage of documents that contain one item. Relative frequency thus has the same denotation as normalized frequency, and relative DOCF means the same as normalized range in AntConc. ARF and ALDF are two modified versions of frequency.

SE has the function of automatic tagging, so both word *n*-grams and POS *n*-grams can be generated. It has two additional functions: access to big data and the automatic segmentation of Chinese texts. When SE generates a keyword list or key *n*-grams, it can utilize a larger corpus directly searched from the Internet; AntConc requires the uploading of a reference corpus by users. Chinese is a highly contextual language, so there are no distinct boundaries between words. Therefore, text segmentation is necessary before data processing (Liu et al. 2022; Wu et al. 2021). Chinese texts are automatically segmented by SE before data processing. AntConc can collaborate with other programs to segment Chinese texts before data processing. Some of these well-known programs include NLPIR developed by CAS,[3] THULAC by Tsinghua University,[4] The Stanford Parser (Levy and Manning

2003) and CorpusWordParser developed by Professor Xiao Han. The last two can be downloaded on the website of Corpus Research Group of Beijing Foreign Studies University.[5]

4.2.2 Corpus-based studies of LBs for translation universals

Baker (1993) put forward the influential theory of translation universals, highlighting the position of translation as a distinct language phenomenon. According to Baker, linguistic characteristics in translation do not occur by chance. They are caused in and by the process of translation. These linguistic features can be manifested in corpus-based comparative studies with non-translated texts. The best-known translation universals include explicitation, simplification, normalization, sanitization, leveling out (or convergence) and shining through, referred to as four-plus-one universals (Dayter 2019; Liu et al. 2022; Xiao 2010).

Mauranen (2004) called for attention to untypical collocation patterns to study translation universals. Mauranen suggested that translated language is influenced by the source text languages due to bilingual processing, resulting in linguistic variations. These linguistic variations exhibit similar patterns across language pairs, indicating that language interference can manifest itself as translation universals. Mauranen formed the hypothesis of untypical collocations, proposing that unique collocation patterns can be identified to distinguish translated texts from non-translated texts.

There has been more research on translation universals using multiword sequences in written translation than in interpreting. In written translation, some studies have identified explicitation in translated texts (Baker 2004; Kruger 2012; Lee 2013; Li and Halverson 2020; Plevoets and Defrancq 2018; Xiao 2011). Some have identified normalization (Kenny 2001; Lewandowska-Tomaszczyk 2012). Xiao (2011) has conducted research on LBs for translation universals in the Chinese context. Xiao concludes that a higher frequency and a wider coverage of LBs indicate a more prominent level of formulaicity in translation. Additionally, Xiao asserts that unique LBs in translated Chinese exhibit specific grammatical preferences. A group of studies have attached importance to grammatical features of LBs. Mauranen (2004) emphasizes that lexical and grammatical distributions indicate atypical collocations in translation. A group of studies have utilized word *n*-grams and POS *n*-grams to differentiate translated from non-translated texts (Alexander Avner et al. 2016; Baroni and Bernardini 2006; Liu et al. 2022).

The studies to identify "interpretese" through the lens of multiword sequences are significantly fewer (Dayter 2019; Setton 2011). Dayter (2019) stands out prominently in this dearth. Inspired by Bernardini (2015) for written translation, Dayter extended the analysis of POS *n*-grams to simultaneous interpreting. The study acknowledges two approaches to identify LBs: the "window-based" and "syntax-based" methods (Dayter 2019, 71). The former focuses on word LBs while the latter examines POS LBs. Dayter utilized the scheme proposed by Bernardini, which includes 13 POS chains to categorize LBs. Additionally, Dayter highlighted the collocativity in analyzing features of LBs. Collocativity encompasses frequency, range and association strength between words. Results suggest that overall, interpreted speech utilizes more frequent and associative word combinations.

A synthesis of CBTS on LBs in comparing translated and non-translated texts suggests that LBs analysis can offer insights into translated language as a distinct code. Specifically, two implications are obtained from this chapter. First, statistical indices including frequency, dispersion and associations are useful; second, word and POS LBs can be utilized to yield the statistical indices.

4.2.3 Methodology-oriented studies on Sketch Engine and AntConc

Methodology-oriented studies for corpus linguistics refer to studies for advancing methodologies, including CTs. There is a lack of empirical studies in comparing CTs. Anthony (2013), Hardie (2012) and Li and Guo (2016) have introduced development of CTs in four generations. Anthony (2013) proposed that disagreements in corpus linguistics regarding the appropriate size of a corpus and the role of annotations can be resolved by the advancement of CTs. Anthony mentioned the open-source and modular fashion of CTs, and the application of large-scale data and advanced computational algorithms. AntWebConc, a CT as an improvement on AntConc, is under development to become one CT in the fourth generation. Hardie (2012, 381) emphasized the balance between "power" and "usability" in the fourth generation of CTs. The fourth generation can be BNCweb based on British National Corpus or CQPweb, the latter considering flexibility by allowing users more available access to corpus data and integration of their own data than the former. SE and Wmatrix are two CQPweb CTs. Li and Guo (2016) focused the discussion of four generations in studying Chinese collocations and demonstrated how to apply AntConc to identify collocations in Chinese corpora, in which manual checking is essential to delete collocations high in frequency but without psychological validity. The three studies contribute to the discussion on SE and AntConc. Only Li and Guo provide empirical data in application. They have yet to attempt the comparison of CTs.

Afanasev (2020) tried the comparison in the study of ancient Greek. Afanasev stated that AntConc and SE are easy to use. They have identical functions although SE has more functions and can do POS tagging. Neither AntConc nor SE is open-source. They cannot be applied in the study of ancient Greek. Researchers must depend on additional software for ancient Greek, which requires programming skills.

One methodology-oriented study is done in CBTS to apply SE to analyze Chinese LBs (Wu et al. 2021). They explored effective ways to identify Chinese LBs so they can summarize primary functions fulfilled by these LBs in international political conferences. They came up with four steps. First, lists of LBs are generated at the frequency threshold; Second, manual checking is needed to exclude the overlapping and topic-specific instances; Third, manual checking of the concordance is needed to identify the remaining LBs that are missed but qualified. Last, a one-gram list is generated to discover overlooked patterns. The differences between SE and AntConc in studying Chinese were mentioned in this study. AntConc depends on additional software like the Chinese Lexical Analysis System (Xiao and Hu 2015) for text segmentation, whereas there is no such need when using SE. Overall, studies focusing on CTs are scarce in CBTS, not to mention empirical ones.

Studies focusing on CTs in CBTS are methodology-oriented, as proposed by Granger and Lefer (2022), who classify CBTS into three main groups: methodological-theoretical, empirical and applied. Methodological-oriented studies are a subfield of methodological-theoretical studies, responsible for methodology and CTs. CTs play an essential role for researchers in identifying distinctive language features in translation and interpreting: Biber et al. (1998) drew attention to the fact that corpus linguistics relies on computer software to count linguistic patterns; Anthony (2013) claimed that the available research methods in corpus linguistics are primarily determined by the functionality provided by CTs; Additionally, Anthony (2013, 149) emphasized the importance to replicate studies using different CTs: "Corpus linguists become most aware of the tools they use when their studies are replicated with a different tool and the results do not match".

This chapter is methodology-oriented, filling the research gap in CBTS to replicate studies through two different CTs. The results are expected to shed light on how CTs can facilitate CBTS. Two research questions (RQs) are formed. In the applications of SE and AntConc,

Rq1: Can they distinguish interpreted from non-interpreted Chinese with the same statistic indices regarding frequency and range of LBs?

Rq2: Can SE use its unique function of POS and nest *n*-grams for the purpose of discrimination?

4.3 Methodology

4.3.1 Corpus

Interpreted and non-interpreted Chinese texts are collected from United Nations conferences, ensuring comparability in terms of genre, topics and procedures. Interpreted Chinese is the result of English-to-Chinese interpretation by multiple interpreters, while native Chinese is delivered by Chinese leaders. Transcriptions of the conferences are subjected to data cleaning to remove noise, messy markers and inaccurate information. Random deletion of interpreted texts balances the word count with non-interpreted texts to yield similar numbers of LBs. The corpus consists of 79 episodes of interpreted speech, totaling 151,199 words, and 91 episodes of non-interpreted speech, totaling 150,626 words. Normalized frequency and range (relative frequency and DOCF in SE) is used to mitigate the impact of slight variations in total word counts.

4.3.2 Identification of LBs

Chinese texts are segmented prior to the generation of LBs. Segmentation can be automated in SE or done in conjunction with CorpusWordParser in AntConc. CorpusWordParser is user-friendly and does not require programming skills. Random sampling is employed to manually check the accuracy of the segmentation, and overall, the 50 samples demonstrate accurate segmentation by CorpusWordParser.

The identification of LBs in this chapter follows two referenced studies: Huang et al. (2023) for English and Wu et al. (2022) for Chinese LBs. The two studies provide detailed descriptions and convincing explanations of the identification. The process of identifying LBs in this chapter involves six steps.

1 Considering the size of the corpus, the threshold of retrieval is set at a minimum of 15 for frequency and five for dispersion in texts; two- to four-word length LBs are under scrutiny to align with previous studies on Chinese LBs (Xiao 2011; Wu et al. 2021);
2 Context-specific LBs are excluded, categorized into three groups. The first group consists of LBs containing proper nouns. The second group includes relatively specialized phrases and terms, such as "小康社会" (moderately prosperous society) and "一体化进程" (integration process). The third group refers to LBs used for addressing and procedural purposes, like "主席先生" (Mr. Chairman) and "谢谢大家" (thank you all). The criterion for determining context-specific LBs relies on the speaker's degrees of freedom in selecting lexicon. Context-specific LBs exhibit limited freedom and are therefore ineffective in revealing tendencies in lexical choices;
3 Overlapping LBs are eliminated. For example, "取得了很" (has made very) is eliminated because an overlapping and more complete bundle "取得了很大" (has made very great) follows; "的 新型 国际 关系" (a new international relationship) appears twice and one is deleted.
4 LBs at the boundary of clauses are eliminated, such as "了 重要" (Le important) "了积极" (Le positive). "了" in conjunction with the preceding verb indicates that an action has already occurred or been completed. The two adjectives modify the nouns that follow.
5 LBs having little psychological reality are excluded. For example, in "呢 是" (Ne is/are), "呢" is a modal particle, "是" means "is/are" and the combination does not make sense and is excluded.
6 LBs are restored for syntactic and semantic integrity. For example, "年的" (of years) identified as a two-unit LB is restored as a three-unit LB "(数字) + 年 的" (of + number + years). The restore is abandoned if the restored LB exceeds four units.

The researcher has acknowledged that manual examining concordance lines, in both SE and AntConc, is necessary to eliminate the unqualified LBs and to restore LBs. Concordance lines are used when the researcher has difficulty in making decisions.

4.3.3 Procedures

After corpus construction and identification of word LBs in SE and AntConc. Mann-Whitney U tests on Jamovi (R Core Team 2021) are done for potential differences regarding frequency (raw, relative and normalized frequency) and dispersion (raw, relative and normalized range) in the first step. A qualitative analysis is then conducted, focusing on common LBs between interpreted and non-interpreted Chinese, as well as unique LBs identified by the two CTs, respectively.

Then, POS n-grams are obtained in SE, and Mann-Whitney U tests are done for differences;

Lastly, nest n-grams are obtained in SE. These nest LBs undergo selection using the same criteria as regular word LBs analyzed in the first step. Mann-Whitney U tests are used again for differences.

4.4 Results and discussion

Section 4.4.1 presents the results for the first RQ, which is to distinguish between interpreted and non-interpreted Chinese based on the frequency and range of regular word LBs by the two CTs. Section 4.4.2 is for the second RQ, showing whether the discrimination can be made by POS and nest n-grams in SE.

4.4.1 Discrimination based on frequency and range

In SE, 2,515 LBs are generated. 1986 LBs are identified using the selection criteria, 1,108 for interpreted and 878 for non-interpreted speech. In AntConc, surprisingly fewer LBs are generated. There are initially 477 LBs, and finally 318 remain, 168 for interpreted and 150 for non-interpreted. Table 4.1 shows the results of SE for the comparison of interpreted and non-interpreted Chinese. Significant differences exist in all categories, indicating that interpreted and native Chinese use LBs differently in terms of frequency and dispersion ($p < .05$).

Figures 4.3 and 4.4 visualize the results of relative frequency and relative DOCF. They show that interpreted Chinese use LBs more frequently and more widely than native Chinese.

Table 4.2 shows the results of AntConc. Unlike SE, AntConc can only discriminate between native interpreted Chinese by range ($p < .05$). Figure 4.5 shows that LBs are more widely distributed in interpreted Chinese than in native Chinese, which is consistent with the findings regarding dispersion in SE.

TABLE 4.1 Results of the Mann-Whitney test in native and interpreted Chinese in SE

		Statistic	*p*
Frequency	Mann-Whitney *U*	454,892	.013*
Relative frequency	Mann-Whitney *U*	430,405	<.001*
DOCF	Mann-Whitney *U*	426,794	<.001*
Relative DOCF	Mann-Whitney *U*	338,316	<.001*
ARF	Mann-Whitney *U*	430,654	<.001*
ALDF	Mann-Whitney *U*	428,358	<.001*

Note: *p < .05

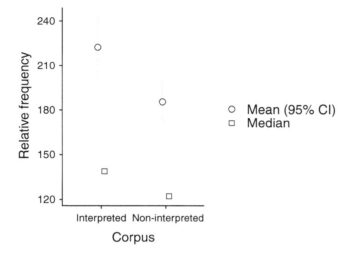

FIGURE 4.3   Results of relative frequency in SE.

The above findings align with previous studies (Dayter 2019; Xiao 2011), high-lighting that interpreted speech relies more on frequent and diverse LBs. Expand-ing upon Xiao's study on Chinese LBs in written translation, this chapter examines the context of simultaneous interpreting. Additionally, it builds upon Dayter's findings specific to English by exploring Chinese LBs in simultaneous interpret-ing. Xiao suggests the increased frequency of LBs in translated Chinese is likely a consequence of the translation process, as translators strive to achieve fluency. This is particularly true in simultaneous interpreting, where simultaneous inter-preters must make immediate decisions on the spot while engaged in dual tasks of comprehension and production (Gile 2008; Seeber 2011). Plevoets and Defrancq (2018) has found that interpreters reduce disfluencies, as indicated by fewer "um", when they use LBs. Striking for fluency may be the reason for a higher frequency in interpreted Chinese found here. In addition, Xiao's study considers the influence of source texts, which may explain the widespread use of LBs in translated texts.

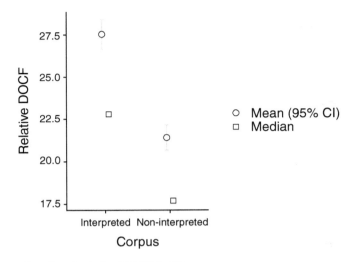

FIGURE 4.4 Results of relative DOCF in SE.

TABLE 4.2 Results of the Mann-Whitney test in native and interpreted Chinese in AntConc

		Statistic	P
Frequency	Mann-Whitney *U*	11,433	.154
Range	Mann-Whitney *U*	10,717	.021*
NormFreq	Mann-Whitney *U*	12,025	.483
NormRange	Mann-Whitney *U*	9,012	<.001*

Note: *p < .05

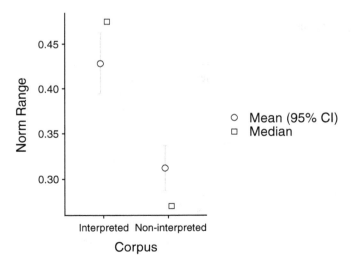

FIGURE 4.5 Results of normalized range in AntConc.

As source text influence is also present in simultaneous interpreting, it supports the wider dispersion of LBs in interpreted Chinese. This wider dispersion confirms the prevalence of LBs in interpreting (Aston 2018; Henriksen 2007), as they are not limited to specific texts but occur commonly across various contexts.

Furthermore, the findings have established multiword sequences as an aspect of descriptive linguistics for discovering distinctive features in interpreting (Dayter 2019). The complex usage identified by multiword sequences in previous studies (Biel 2018; Li and Halverson 2020; Xiao 2011) is replicated in this chapter. As in these previous studies, this chapter predicts that employing more LBs is a strategy to cope with more cognitive constraints in interpreted speech than in original speech.

Notably, AntConc does not reveal significant differences in frequency. One possible explanation is that AntConc generates far fewer LBs compared to SE (477 vs. 2515). SE and AntConc share 134 LBs in interpreted speech and 112 in non-interpreted speech. Considering that AntConc identifies 168 LBs in interpreted speech and 150 in non-interpreted speech, it can be said that SE covers over two-thirds of the LBs identified by AntConc but provides more. Sufficient sample sizes can facilitate the detection of frequency differences.

The significant discrepancy in the number of generated LBs between the two CTs may stem from two factors. Firstly, they may employ different mathematical and computational algorithms for LB identification. Secondly, SE automatically segments the texts, while AntConc collaborates with CorpusWordParser for this task. The precise reason remains unknown.

The following section presents the results of qualitative analysis to provide more details in the comparisons.

SE identifies 207 common LBs (types) between interpreted and non-interpreted Chinese, whereas AntConc only identifies 11 common LBs. To categorize these 207 common LBs for observation, Table 4.3 categorizes them based on their POS, focusing on the top 10 most frequent POS. As shown in Table 4.3, both interpreted Chinese and native Chinese demonstrate a preference for LBs containing the particle "的" (De). These LBs fall into four groups: "particle + noun", "verb + particle", "noun + particle" and "adjective + particle", totaling 71 types. The most prevalent group is "particle + noun". Chinese is a head-final language and "的" is a mighty category in modern Chinese, widely used in various modifier-head structures where the head can either appear or not appear (Liu 2022). The frequent usage of LBs with "的" is thus expected

Twenty unique most frequent LBs in interpreted Chinese identified by SE are "的 这个", "的 人", "的 时候", "必须 要", "我们 必须", "所有 的", "他们 的", "挑战", "的 工作", "在 我们", "（数字）个 国家", "多 的", "方面 的", "的 人民", "这方面", "它 的", "人 的", "的 行动", "全球 的" and "不 能够".

Twenty unique most frequent LBs in native Chinese identified by SE are "各 国", "世界 经济", "命运 共同体", "和平 稳定", "合作 共赢", "互联 互通", "基 础 设施", "（数字）国 人民", "之 路", "世界 和平", "国际 关系", "共赢 的",

TABLE 4.3 Common LBs in interpreted and non-interpreted speech identified by SE (only for top ten most frequent POS)

POS	Type of LBs	Examples
Particle + noun	33	的 安全、的 成功、的 成果
Verb + particle	19	包容 的、持续 的、带来 的
Noun + particle	12	地区 的、关系 的、国家 的
Noun + noun	7	国际 合作、国际 社会、经济 社会
Adjective + particle	11	和平 的、好 的、共同 的
Adverb +modal verb	6	不 会、不 可、不 能
Adverb + link verb	6	不 是、都 是、就 是
Adverb + verb	6	也 有、都 有、还 有
Adverb + adjective	6	更 多、更 好、最 大
Numerical + noun for time	5	十年、一年、千 年

"全球 经济", "共同 发展", "地区 国家", " (数字) 国 集团", "人类 命运", "中方 愿", "人类 命运 共同体" and "经济 全球化".

Further analysis of the most frequent and unique LBs in interpreted and native Chinese (see above) shows that LBs with "的" are frequently included in the most frequent LBs in interpreted Chinese (14/20), while they are less frequently included in the most frequent LBs in native Chinese (1/20). A majority of the most frequent LBs in non-interpreted speech are technical and formulaic terms that rely heavily on concrete words and get rid of functional words. Examples include "世界 经济" (world economy), "命运 共同体" (community of shared destiny) and "合作 共赢" (win-win cooperation).

Table 4.4 summarizes the common LBs in interpreted and native Chinese identified by AntConc. The numbers in brackets indicate their ranking from high to low frequency. The 12 types of LBs are all covered by SE. The preference for LBs with particle "的" still exists. Half of the 12 types include particle "的".

Twenty unique most frequent LBs in interpreted speech identified by AntConc are "在 这 方面", "更 好 的", "我们 必须 要", "我们 的", "所 做 的", "和平 与 安全", "很 大 的", "可 持续 的", "和平 和 安全", "所 面临 的", "很 好 的", "非常 重要 的", "千 年 发展", "和平 安全 和", "(数字) 年 发展 目标", "气候 变化 的", "的 解决 办法", "一个 新 的", "更 大 的" and "包容性 的". The unique LBs found by AntConc for interpreted Chinese do not overlap with any of those found by SE.

Twenty unique most frequent LBs in non-interpreted speech identified by AntConc are "世界 经济*", "发展 的", "我们 将", "合作 的", "和平 稳定*", "命运 共同体", "的 重要", "合作 共赢*", "经济 增长", "互联 互通*", "产 能", "基础 设施*", "世界 和平*", "的 国际", "共赢 的*", "国际 关系*", "的 合作", "全球 经济*", "共同 发展*" and "和平 与". Seven LBs (with "*" in the end) overlap with those found by SE.

Further analysis of the most frequent and unique LBs in interpreted and non-interpreted Chinese in AntConc (see above) shows that LBs with "的" are

TABLE 4.4 Common LBs in interpreted and non-interpreted speech identified by AntConc

Lexical bundles	Part of speech	Normalized Frequency	
		Interpreted	*Non-interpreted*
的 发展	Particle + noun	4,100.515 (5)	6,472.703 (3)
发展 的	Noun + particle	2,129.114 (10)	6,995.75 (1)
国家 的	Noun + particle	4,599.937 (4)	2,876.757 (8)
更 好	Adverb + adjective	2,943.96 (9)	2,255.639 (10)
国际 社会	Noun + noun	3,785.091 (8)	4,805.492 (6)
国际 社会 的	Noun + noun + particle	6,944.444 (2)	4,883.766 (5)
可 持续 的	Modal verb + verb + particle	1,6319.444 (1)	6,251.221 (3)
气候 变化	Noun + noun	3,916.518 (6)	2,549.853 (9)
人民 的	Noun + particle	1,813.689 (11)	1,961.425 (11)
我们 要	Pronoun + modal verb	6,015.038 (3)	6,492.482 (2)
新的	Adjective + particle	4,994.217 (4)	6,113.109 (4)
这一	Pronoun + numerical	3,522.237 (7)	3,595.946 (7)

frequently contained by the most frequent LBs (13/20) in interpreted Chinese, similar to the results in SE. Notably, the LBs with "的" identified by AntConc here play the role of attributes. Of the 13 LBs with "的", 12 modify following nouns or noun phrases, such as "更好的" (better) and "我们的" (our). This tendency is less obvious in unique LBs with "的" in interpreted Chinese identified by SE. In the SE group, seven LBs with "的" are attributes such as "所有的" (all) and "他们的" (their), while there are still four LBs that use "的" to emphasize the nouns or noun phrases being modified, such as "的人" (person/people who), "的时候" (time when) and "的挑战" (the challenge that). The head-final tendency is manifested in the interpreted Chinese here.

Furthermore, the tendency to avoid functional words in native Chinese is less pronounced in AntConc. Functional words are more common in LBs identified by AntConc for native Chinese. Eight LBs among the 20 most frequent unique LBs of native Chinese use functional words. Three functional words are "的", "将" (be going to) and "与" (and). In the SE group, there are only three LBs with three functional words "的", "之" (a synonym of "的") and "愿" (wish).

4.4.2 Discrimination based on POS and nest n-grams in SE

Tables 4.5 and 4.6 show the results of discrimination using the POS and nest *n*-gram functions to generate LBs in SE. The same threshold of at least 15 frequencies is used. SE generates 2,515 word LBs but 2,869 POS *n*-grams, 1,553 for interpreted and 1,316 for native Chinese. The nest *n*-gram feature does not reduce the number of LBs. A total of 2,516 nest *n*-grams are generated. After manual selection, 1,999 LBs remain, 1,100 for interpreted and 899 for non-interpreted Chinese.

TABLE 4.5 Results of the Mann-Whitney test in native and interpreted Chinese in SE (POS)

		Statistic	p
Frequency	Mann-Whitney *U*	998,465	.290
Relative frequency	Mann-Whitney *U*	1,020,000	.863
DOCF	Mann-Whitney *U*	961,283	.006*
Relative DOCF	Mann-Whitney *U*	923,605	<.001*
ARF	Mann-Whitney *U*	1,000,000	.398
ALDF	Mann-Whitney *U*	1,010,000	.484

Note: *p < .05

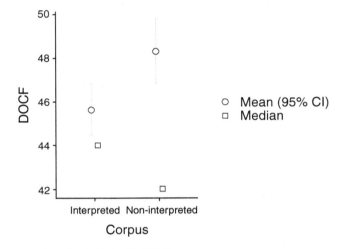

FIGURE 4.6   Results of DOCF in SE (POS).

In contrast to the findings based on word LBs in SE, where significant differences are found for frequency and dispersion in texts, significant differences exist only for dispersion based on POS LBs ($p < .05$). Even this finding remains puzzling. Figure 4.6 for raw DOCE indicates that native Chinese use more widespread POS LBs than interpreted speech, contradicting findings based on word LBs in SE. Surprisingly, in terms of relative DOCF, shown in Figure 4.7, it is interpreted speech that uses more widespread POS LBs than non-interpreted speech.

Table 4.6 shows the result of discrimination based on nest *n*-grams in SE. The results are consistent with those based on regular word LBs in SE (see Table 4.1). There are significant differences in frequency and dispersion. The results remain stable between raw and relative frequency, as indicated by Figures 4.8 and 4.9, and between DOCF and relative DOCF, as shown by Figures 4.10 and 4.11. Interpreted

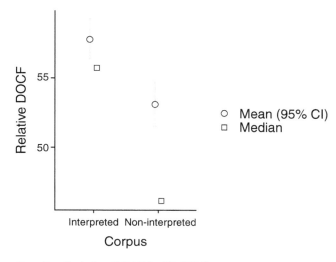

FIGURE 4.7 Results of relative DOCF in SE (POS).

TABLE 4.6 Results of the Mann-Whitney test in native and interpreted Chinese in SE (nest *n*-grams)

		Statistic	p
Frequency	Mann-Whitney *U*	455,810	.003*
Relative frequency	Mann-Whitney *U*	430,747	<.001*
DOCF	Mann-Whitney *U*	430,826	<.001*
Relative DOCF	Mann-Whitney *U*	341,226	<.001*
ARF	Mann-Whitney *U*	431,651	<.001*
ALDF	Mann-Whitney *U*	429,538	<.001*

Note: *p < .05

Chinese rely on a higher frequency and a wider distribution of LBs than non-interpreted Chinese, suggesting that complexification takes place in the interpreting process.

When comparing the results of the unique functions (POS and nest *n*-grams) with regular word LBs in SE, and between the two unique functions themselves, it seems questionable to rely solely on POS LBs without manual filtering. The word meanings of POS LBs remain unknown, making it impossible to apply the criteria for filtering out unqualified LBs. The results of POS LBs differ significantly from those based on LBs obtained through manual selection, either regular LBs or nest LBs. It is difficult to draw a conclusion regarding the existence of translation universals in interpreted Chinese using POS LBs.

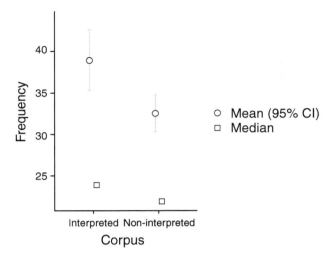

FIGURE 4.8 Results of frequency in SE (nest *n*-grams).

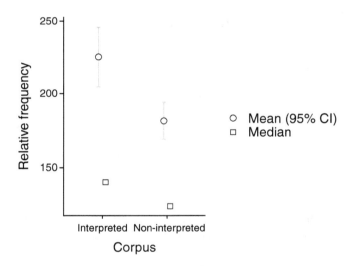

FIGURE 4.9 Results of relative frequency in SE (nest *n*-grams).

When comparing the applications of regular LBs and nest *n*-grams, the results remain consistent. The two approaches generate a similar number of LBs and thus require a similar amount of research time and effort for manual concordance checking and filtering. Both approaches can be used to distinguish between interpreted and non-interpreted Chinese.

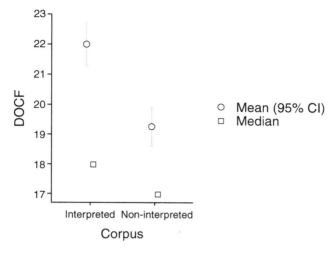

FIGURE 4.10   Results of DOCF in SE (nest *n*-grams).

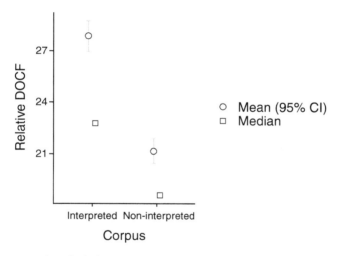

FIGURE 4.11   Results of relative DOCF in SE (nest *n*-grams).

4.5 Conclusion

In this chapter, the latest versions of two CTs, SE and AntConc, are used to distinguish between interpreted and non-interpreted Chinese through the lens of three types of LBs in terms of their frequency and distribution in texts. The two CTs are similarly convenient to use. Both have similar running speeds and moderate system requirements. Although AntConc cannot perform Chinese text segmentation automatically, the problem can be solved by cooperating with another text segmentation tool CorpusWordParser.

SE provided significantly more LBs, covering most of those identified by AntConc. Consequently, it takes more time and effort to process the data. However, a sufficient amount of data is needed to distinguish between interpreted and non-interpreted Chinese on the basis of frequency. In general, both CTs effectively detect a more complex usage of LBs in interpreted Chinese, contributing to the inclusion of a non-European language in the study of translation universals. Qualitative analysis using highly frequent LBs provided by the two CTs reveals certain tendencies. For example, LBs containing "的" (de) are preferred. SE shows a stronger preference for informative concrete words in native Chinese.

The results based on part-of-speech LBs in SE differ from any other approach used. This discrepancy could be due to the inability to manually filter out unneeded LBs or to the specific characteristics of such LBs. The effort required to analyze regular LBs and nest n-grams in SE is similar, and the results are identical.

CT provides substantial convenience for comparing language features between interpreted and non-interpreted speech. However, further caution is needed in understanding how to use CTs effectively and how to interpret the results, due to variations resulting from the operating principles and computational algorithms of different CTs. Similar attempts to compare different CTs when they are applied for the same research purposes are needed, in different contexts by incorporating more factors and languages.

Notes

1 https://www.laurenceanthony.net/.
2 https://www.sketchengine.eu/.
3 http://ictclas.nlpir.org/nlpir/.
4 https://github.com/thunlp/THULAC.
5 https://corpus.bfsu.edu.cn/index.htm.

References

Afanasev, Ilia. 2020. "A Corpus-Based Approach in Archaeolinguistics." *Journal of Applied Linguistics and Lexicography* 2(2): 147–59. https://doi.org/10.33910/2687-0215-2020-2-2-147-159.

Alexander Avner, Ehud, Noam Ordan, and Shuly Wintner. 2016. "Identifying Translationese at the Word and Sub-Word Level". *Digital Scholarship in the Humanities* 31(1): 30–54. https://doi.org/10.1093/llc/fqu047.

Anthony, Laurence. 2005. *AntConc: A Learner and Classroom Friendly, Multi-platform Corpus Analysis Toolkit.* IWLeL 2004: An Interactive Workshop on Language e-Learning, Waseda University, Tokyo.

Anthony, Laurence. 2006. "Developing a Freeware, Multiplatform Corpus Analysis Toolkit for the Technical Writing Classroom." *IEEE Transactions on Professional Communication* 49(3): 275–86. https://doi.org/10.1109/tpc.2006.880753.

Anthony, Laurence. 2013. "A Critical Look at Software Tools in Corpus Linguistics." *Linguistic Research* 30(2): 141–61.

Anthony, Laurence. 2019. "Tools and Strategies for Data-driven Learning (DDL) in the EAP Writing Classroom". In *Specialised English: New Directions in ESP and EAP Research and Practice*, edited by Ken Hylan and Lillian L. C. Wong, 179–94. London/New York: Routledge.

Aston, Guy. 2018. "Acquiring the Language of Interpreters: A Corpus-Based Approach". In *Making Way in Corpus-Based Interpreting Studies*, edited by Mariachiara Russo, Claudio Bendazzoli, and Bart Defrancq, 83–96. Singapore: Springer.

Baker, Mona. 1993. "Corpus Linguistics and Translation Studies: Implications and Applications". In *Text and Technology: In Honour of John Sinclair*, edited by Mona Baker, Gill Francis, and Elena Tognini-Bonelli, 233–50. Amsterdam/Philadelphia, PA: John Benjamins Publishing.

Baker, Mona. 2004. "A Corpus-Based View of Similarity and Difference in Translation". *International Journal of Corpus Linguistics* 9(2): 167–93. https://doi.org/10.1075/ijcl.9.2.02bak.

Barlow, Michael. 2000. *MonoConc Pro*. Houston: Athelstan.

Baroni, Marco, and Silvia Bernardini. 2006. "A New Approach to the Study of Translationese: Machine-Learning the Difference between Original and Translated Text." *Literary and Linguistic Computing* 21(3): 259–74. https://doi.org/10.1093/llc/fqi039.

Bernardini, Silvia. 2015. "Translation". In *The Cambridge Handbook of English Corpus Linguistics*, edited by Douglas Biber and Randi Reppen, 515–36. Cambridge, UK: Cambridge University Press.

Biber, Douglas, Susan Conrad, and Viviana Cortes. 2004. "If You Look at …: Lexical Bundles in University Teaching and Textbooks". *Applied Linguistics* 25(3): 371–405. https://doi.org/10.1093/applin/25.3.371.

Biber, Douglas, Susan Conrad, and Randi Reppen. 1998. *Corpus Linguistics*. Cambridge, UK: Cambridge University Press.

Biel, Łucja. 2018. "Lexical Bundles in EU Law: The Impact of Translation Process on the Patterning of Legal Language." In *Phraseology in Legal and Institutional Settings: A Corpus-Based Interdisciplinary Perspective*, edited by Stanisław Goźdź-Roszkowski and Gianluca Pontrandolfo, 11–26. London/New York: Routledge.

Biel, Lucja, Dariusz Kozbial, and Katarzyna Wasilewska. 2019. "The Formulaicity of Translations Across EU Institutional Genres A Corpus-Driven Analysis of Lexical Bundles in Translated and Non-Translated language". *Translational Spaces* 8(1): 67–92. https://doi.org/10.1075/ts.00013.bie.

Cheung, Andrew Kay Fan, Kanglong Liu, and Riccardo Moratto. 2023. "East Asian Perspectives on Corpus-based Interpreting Studies". In *Corpora in Interpreting Studies: East Asian Perspectives*, edited by Andrew Kay Fan Cheung, Kanglong Liu, and Riccardo Moratto. London/New York: Routledge. https://doi.org/10.4324/9781003377931-1.

Dayter, Daria. 2019. "Collocations in Non-Interpreted and Simultaneously Interpreted English: A Corpus Study." In *New Empirical Perspectives on Translation and Interpreting*, edited by Lore Vandevoorde, Joke Daems, and Bart Defrancqpp, 67–91. London/New York: Routledge.

Frawley, William. 1984. "Prolegomenon to a Theory of Translation". In *Translation: Literary, Linguistic, and Philosophical Perspectives*, edited by William Frawley, 159–75. Newark: University of Delaware Press.

Fu, Rongbo, and Kefei Wang. 2022. "Hedging in Interpreted and Spontaneous Speeches: A Comparative Study of Chinese and American Political Press Briefings." *Text & Talk* 42(2): 153–75. https://doi.org/https://doi.org/10.1515/text-2019-0290.

Gile, Daniel. 2008. "Conference Interpreting, Historical and Cognitive Perspectives." In *Routledge Encyclopedia of Translation Studies*, edited by Mona Baker and Gabriela Saldanha, 51–56. London/New York: Routledge.

Götz, Sandra. 2021. "Analyzing a Learner Corpus with a Concordancer." In *The Routledge Handbook of Second Language Acquisition and Corpora*, edited by Nicole Tracy-Ventura and Magali Paquo, 68–89. New York: Routledge. https://doi.org/10.4324/9781351137904-8.

Granger, Sylviane, and Marie-Aude Lefer. 2022. "Corpus-Based Translation and Interpreting Studies: A Forward-Looking Review." In *Extending the Scope of Corpus-Based Translation Studies*, edited by Sylviane Granger and Marie-Aude Lefer, 13–41. London: Bloomsbury. https://doi.org/10.5040/9781350143289.0007.

Hardie, Andrew. 2012. "CQPweb – Combining Power, Flexibility and Usability in a Corpus Analysis Tool." *International Journal of Corpus Linguistics* 17(3): 380–409. https://doi.org/10.1075/ijcl.17.3.04har.

He, Anping. 2015. "Corpus Pedagogic Processing of Phraseology for EFL Teaching: A Case of Implementation." In *Corpus Linguistics in Chinese Contexts*, edited by Bin Zou, Simon Smith, and Michael Hoey, 98–113. New York: Palgrave Macmillan.

Henriksen, Line. 2007. "The Song in the Booth: Formulaic Interpreting and Oral Textualisation". *Interpreting: International Journal of Research and Practice in Interpreting* 9(1): 1–20. https://doi.org/10.1075/intp.9.1.02hen.

Huang, Chu-Ren, Adam Kilgarriff, Yiching Wu, Chih-Ming Chiu, Simon Smith, Pavel Rychlý, Ming-Hong Bai, and Keh-Jiann Chen. 2005. "Chinese Sketch Engine and the Extraction of Grammatical Collocations." *Proceedings of the Fourth SIGHAN Workshop on Chinese Language Processing*, 48–55. Jeju Island, Korea.

Huang, Dan Feng, Fang Li, and, Hang Guo. 2023. "Chunking in Simultaneous Interpreting: The Impact of Task Complexity and Translation Directionality on Lexical Bundles." *Frontiers in Psychology* 14:1252238. https://doi.org/10.3389/fpsyg.2023.1252238.

Jones, Roderick. 1998. *Conference Interpreting Explained*. Manchester: St. Jerome Pub.

Kaye, Geoffrey. 1990. "A Corpus Builder and Real-Time Concordance Browser for an IBM PC." *Theory and Practice in Corpus Linguistics* 4: 137–61.

Kenny, Dorothy. 2001. *Lexis and Creativity in Translation: A Corpus-Based Study*. Manchester/Northampton, MA: St. Jerome Pub.

Kilgarriff, Adam, Vit Baisa, Jan Bušta, Miloš Jakubíček, Vojtěch Kovář, Jan Michelfeit, Pavel Rychlý, and Vít Suchomel. 2014. "The Sketch Engine: Ten Years On". *Lexicography (Berlin)* 1(1): 7–36. https://doi.org/10.1007/s40607-014-0009-9.

Kilgarriff, Adam, Fredrik Marcowitz, Simon Smith, and James Thomas. 2015. "Corpora and Language Learning With the Sketch Engine and SKELL." *Revue française de linguistique appliquée* 20(1): 61–80. https://doi.org/10.3917/rfla.201.0061.

Kruger, Haidee. 2012. "A Corpus-Based Study of the Mediation Effect in Translated and Edited Language". *Target: International Journal of Translation Studies* 24(2): 355–88. https://doi.org/10.1075/target.24.2.07kru.

Lee, Changsoo. 2013. "Using Lexical Bundle Analysis as Discovery Tool for Corpus-Based Translation Research". *Perspectives, Studies in Translatology* 21(3): 378–95. https://doi.org/10.1080/0907676x.2012.657655.

Levy, Roger, and Christopher Manning. 2003. "Is it Harder to Parse Chinese, or the Chinese Treebank?" In *The 41st Annual Meeting of the Association for Computational Linguistics*, 439–46. Sapporo, Japan.

Lewandowska-Tomaszczyk, Barbara. 2012. "Explicit and Tacit." In *Quantitative Methods in Corpus-Based Translation Studies: A Practical Guide to Descriptive Translation Research*, edited by Michael P. Oakes and Meng Ji, 51: 3–34. Amsterdam/Philadelphia, PA: John Benjamins Publishing Company.

Li, Shihai, and Shufen Guo. 2016. "Collocation Analysis Tools for Chinese Collocation Studies." *Journal of Technology and Chinese Language Teaching* 7(1): 56–77.

Li, Yang, and Sandra Halverson. 2020. "A Corpus-Based Exploration into Lexical Bundles in Interpreting." *Across Languages and Cultures* 21(1): 1–22. https://doi.org/10.1556/084.2020.00001.

Liu, Xueying 刘雪莹. 2022. "The Particle "Zhi" in the Modern Chinese and Its Relation With "De"" 现代汉语中的助词"之"及其与"的"的关系. 暨南学报 （哲学社会科学版） [*Jinan Journal (Philosophy & Social Sciences)*] 287(12): 81–89.

Liu, Yi. 2023. "An Investigation of Lexical Bundles in L1 Speeches and L2 Interpreted Languages: A Corpus-Based Study." PhD dissertation, Hong Kong Polytechnic University.

Liu, Kanglong, and Muhamand Afzaal. 2021. "Syntactic Complexity in Translated and Non-Translated Texts: A Corpus-Based Study of Simplification." *PLoS One* 16(6): e0253454. https://doi.org/10.1371/journal.pone.0253454.

Liu, Yi, Andrew Kay Fan Cheung, and Kanglong Liu. 2023. "Syntactic Complexity of Interpreted, L2 and L1 Speech: A Constrained Language Perspective." *Lingua* 286: 103509. https://doi.org/10.1016/j.lingua.2023.103509.

Liu, Kanglong, Rongguang Ye, Zhongzhu Liu, and Rongye Ye. 2022. "Entropy-Based Discrimination between Translated Chinese and Original Chinese Using Data Mining Techniques." *PLoS One* 17(3): e0265633. https://doi.org/10.1371/journal.pone.0265633.

Mauranen, Anna. 2004. "Corpora, Universals and Interference." In *Translation Universals: Do They Exist*, edited by Pekka Kujamäki and Anna Mauranen, 65–82. Amsterdam/Philadelphia, PA: John Benjamins Publishing Company.

Mauranen, Anna, and Pekka Kujamäki. 2004. *Translation Universals: Do They Exist?* Amsterdam/Philadelphia, PA: John Benjamins Publishing Company.

Pearce, Michael. 2008. "Investigating the Collocational Behaviour of MAN and WOMAN in the BNC Using Sketch Engine." *Corpora* 3(1): 1–29. https://doi.org/10.3366/E174950 320800004X.

Plevoets, Koen, and Bart Defrancq. 2018. "The Cognitive Load of Interpreters in the European Parliament A Corpus-Based Study of Predictors for the Disfluency Uh(m)." *Interpreting: International Journal of Research and Practice in Interpreting* 20(1): 1–28. https://doi.org/10.1075/intp.00001.ple.

Rayson, Paul. 2008. "From Key Words to Key Semantic Domains." *International Journal of Corpus Linguistics* 13(4): 519–49. https://doi.org/10.1075/ijcl.13.4.06ray.

R Core Team. 2021. R: A Language and Environment for Statistical Computing. (Version 4.1) [Computer software]. https://cran.r-project.org.

Reed, Alan. 1978. *CLOC User Manual*. UK: University of Birmingham.

Scott, Michael. 2001. "Comparing Corpora and Identifying Key Words, Collocations, and Frequency Distributions through the WordSmith Tools Suite of Computer Programs." In *Small Corpus Studies and ELT: Theory and Practice*, edited by Mohsen Ghadessy, Robert L. Roseberry, and Alex Henry, 47–67. Amsterdam/Philadelphia, PA: John Benjamins Publishing Company.

Seeber, Kilian. 2011. "Cognitive Load in Simultaneous Interpreting: Existing Theories – New Models." *Interpreting: International Journal of Research and Practice in Interpreting* 13(2):176–204. https://doi.org/10.1075/intp.13.2.02see.

Setton, Robin. 2011. "Corpus-Based Interpreting Studies (CIS): Overview and Prospects." In *Corpus-Based Translation Studies: Research and Applications*, edited by Alet Kruger, Kim Wallmach, and Jeremy Munday, 33–75. London: Continuum.

Shlesinger, Miriam. 2008. "Towards a Definition of Interpretese: An Intermodal, Corpus-Based Study." In *Efforts and Models in Interpreting and Translation Research a Tribute to Daniel Gile*, edited by Hansen Gyde, Chesterman Andrew, and Gerzymisch-Arbogast Heidrun, 237–53. Amsterdam/Philadelphia, PA: John Benjamins Publishing Company.

Stubbs, Michael. 2007. "An Example of Frequent English Phraseology: Distributions, Structures and Functions." In *Corpus Linguistics 25 Years On*, edited by Roberta Facchinetti, 89–105. Amsterdam/New York: Brill.

Su, Yanfang, Kanglong Liu, and Andrew Kay Fan Cheung. 2023. "Epistemic Modality in Translated and Non-Translated English Court Judgments of Hong Kong: A Corpus-Based Study." *The Journal of Specialised Translation* 40: 56–80.

Thomas, James. 2017. *Discovering English with Sketch Engine: A Corpus-Based Approach to Language Exploration*. Brno, Czech Republic: Versatile.

Wang, Shan, and Chu-Ren Huang. 2013. "Apply Chinese Word Sketch Engine to Facilitate Lexicography." *Lexicography and Dictionaries in the Information Age: Selected Papers from the 8th ASIALEX International Conference*, 285–92. Bali, Indonesia.

Wu, Yinyin. 2021. "Lexical Bundles in English EU Parliamentary Discourse-Variation across Interpreted, Translated, and Spoken Registers." *Compilation & Translation Review* 14(2): 188–206. https://doi.org/10.1080/1750399x.2018.1451952.

Wu, Baimei, Andrew Kay Fan Cheung, and Jie Xing. 2021. "Learning Chinese Political Formulaic Phraseology from a Self-Built Bilingual United Nations Security Council Corpus: A Pilot Study." *Babel* 67: 500–21. https://doi.org/10.1075/babel.00233.wu.

Xiao, Richard. 2011. "Word Clusters and Reformulation Markers in Chinese and English Implications for Translation Universal Hypotheses." *Languages in Contrast* 11(2): 145–71. https://doi.org/10.1075/lic.11.2.01xia.

Xiao, Richard, Lianzhen He, and Ming Yue. 2010. "In Pursuit of the Third Code: Using the ZJU Corpus of Translational Chinese in Translation Studies." In *Using Corpora in Contrastive and Translation Studies*, edited by Richard Xiao, 182–214. London: Cambridge Scholars Publishing.

Xiao, Richard, and Xianyao Hu. 2015. *Corpus-Based Studies of Translational Chinese in English-Chinese Translation*. Heidelberg: Springer.

5

A STUDY ON THE IMPACT AND INSIGHTS OF TECHNOLOGY ON INTERPRETING EDUCATION

Zhi Li

5.1 Introduction

The development of interpreting-related technologies is closely linked to advancements in audio and video transmission, as well as natural language processing technologies. These technologies have transformed interpreting from a language service (such as consecutive and simultaneous interpreting) into a specialized skill and profession, from the voice transmission devices used in consecutive interpreting to simultaneous interpreting equipment and from automatic speech recognition plus machine translation for simultaneous interpreting to cloud computing plus online conferencing for remote interpreting services.

In the professionalization of interpreting, interpreting education and training are particularly important. While machine simultaneous interpreting by Baidu, DeepL, Google and Tencent around 2021 might not have posed a significant threat to interpreting educators and interpreters, the recent rise of generative artificial intelligence (AI), represented by ChatGPT, undeniably presents substantial challenges and impacts on the future of interpreting education and industry development.

In the early stages of its development, technology failed to fully meet the requirements of interpreters and interpreting education due to its constraints and imperfect functions. Consequently, many interpreters and educators reluctantly embraced technology, believing it to be inadequate and potentially detrimental to interpreting quality. However, as technology progressed and improved, interpreters and educators began reevaluating its merits. The recent advent of generative AI has profoundly impacted traditional educational paradigms, diminishing the emphasis on factual knowledge and memorization while steadily eroding language barriers among widely spoken languages (Feng and Zhang 2024). According to the "Record and Approval Results of Undergraduate Programs of Regular

DOI: 10.4324/9781003597711-6

Higher Education Institutions" from 2018 to 2022 issued by China's Ministry of Education, over 100 universities have ceased enrollment in certain foreign language majors. This phenomenon is also evident in other countries globally, including South Korea, the United States and the United Kingdom.

5.2 Interpreting education and technology

Technologies influencing interpreter training can be divided into two categories. One category involves technologies that need learning as part of the teaching content, which interpreters (learners) need to learn for practical interpreting practice. The other category includes technologies relating to the teaching methods and modes, to enhance teaching effectiveness and quality. The integration of AI technology with these two categories has made previously difficult tasks possible and diversified their roles in interpreter training.

Interpreter training contents include not only traditional content such as language proficiency and interpreting skills but also technologies used in professional interpreting practice. To ensure the smooth conduct of interpreting courses and training, learners must flexibly apply various technologies before, during and after classes. Wang and Li (2022) categorized the interpreting technologies that make a difference in practice, including interpreting devices, computer applications, information retrieval, terminology management, computer-aided translation (CAT) and other interpreting technologies (such as speech recognition). Most of these technologies are used by interpreters (learners) in their professional practice, but they also include some technologies used in interpreting education (such as interpreting teaching platforms) and research (such as interpreting corpora). In training, students need to master the technical content that interpreters use in their practice.

Before classes, students need to interpreting preparations, similarly to that of interpreters' conduct. The preparation includes glossary, parallel sentence pairs and background information. The technologies used and learned during this stage include computer applications, information retrieval, terminology management and CAT.

During the class, students use different technologies depending on the type of interpreting they learn. For consecutive interpreting, a language lab usually suffices. However, for simultaneous interpreting, remote interpreting and telephonic interpreting, students probably learn how to use the corresponding equipment, including simultaneous interpreting systems and remote interpreting devices.

Teachers can utilize various assisting teaching technologies, such as blended learning through MOOC and SPOC platforms, immersive learning platforms based on virtual reality (VR), online interpreting resources and customized courses through large language models. Additionally, the above teaching assistances can also evaluate student learning outcomes with AI.

After class, students need to continue reinforcing and consolidating their interpreting skills through review and assignments. Repeated practice and using technology learned in class are effective methods for mastering the application of technologies. Students use computer technology to review and reorganize the learning materials, evaluate and analyze peer performance, transcript audio and video materials and utilize interpreting training platforms to further enhance their proficiency.

5.3   Integration of technology with interpreting training content

5.3.1   Information retrieval

Conducting information retrieval, students can use both search engines combined with advanced search syntax and online corpora and dictionaries, and professional databases as well. Additionally, they can quickly obtain information by using generative AI (such as ChatGPT). Generally, using syntax like "site:" and "filetype": in search engines (such as Google) can quickly locate certain information, but interpreters still need to browse, analyze and summarize it, which can be time-consuming. In contrast, Generative AI has significant advantages in the speed of organization and summarization and large searching scope of information retrieval. By inputting clear keyword prompts into the dialogue box and continuously querying related information, learners can quickly acquire useful information or find clues to where helpful information can be found. Xu et al. (2024) utilized GPT-4 to set prompts for finding background information and important concepts.

5.3.1.1   Prompt 1

"User: Based on the excerpt provided from *the interview transcript*, can you determine the overarching subject of the discussion? Also, please list a few specific topics mentioned in this excerpt. **"Paste your text here"**
    GPT: […]"[1]
    Prompt 1 involves conducting background analysis, querying and organizing through existing materials. The italicized part (added by the paper authors) "interview transcript" can be replaced with other content to indicate the material type. At the end, you can paste the text to be analyzed. When there are no reference materials available and online searches are needed, ChatGPT can provide a clear direction.

5.3.1.2   Prompt 2

"User: Could you create a study sheet that introduces key technical concepts from this conversation, as well as those closely related to the topic that might appear in similar discussions? Aim for concise and simple explanations suitable for a general audience with basic or intermediate prior knowledge of *the subject*. Please focus on concepts that are essential for understanding the overarching topic.

GPT: [...]"[2]

When there are no reference materials available and searches are conducted based on themes, learners need to sift through a large amount of information on search engines, which is demanding and sometimes ineffective. With the assistance of ChatGPT, prompts can quickly provide an outline and relevant knowledge on the topic. Similar to Prompt 1, the italicized parts can be replaced as needed.

5.3.2 Terminology management

As for terminology management, interpreting has evolved from using pen and paper for note-taking to employing office software like Excel and Word, and then to specialized terminology tools such as InterpretBank,[3] Multiterm, Intragloss and Termbox of Lingosail.[4] Generative AI can be used for terminology management, such as term extraction and translation. The pen-paper way has already been replaced by subsequent terminology tools under the BYOD (Bring Your Own Device) model. Utilizing specialized terminology management software is available both online and offline. These tools have their own procedures and uploaded text requirements, such as text format, word count limitations and extraction rules, which learners need to learn before using them. Moreover, most of these tools extract terms based on natural language processing and corpus linguistics tokenization principles, so the extracted content may contain more than necessary and requires manual proofreading. The fourth extraction method, is significantly faster and more accurate than the previous three, especially when bilingual texts are provided. Xu et al. (2024) set and tested prompts for quick terminology extraction using GPT-4 and other large language models.

5.3.2.1 Prompt 3[5]

"User: Could you identify the *technical terms* related to the topic of the provided *interview transcript* and translate them into *Chinese*? Please list the terms alongside their Chinese translations in two separate columns of a table." **"Paste your text here"**

In Prompt 3, the italicized parts can be modified or replaced with specific terms learners want to search for. Add professional qualifiers before "technical terms", such as medical, clinical or pediatric, to specify the domain (e.g., medical terms). The text genre to extract from is an interview transcript, but this can be replaced with other types of text, such as a thesis that a speaker needs to read. "Chinese" in the prompt refers to translating the extracted terms into Chinese; both the source and target languages can be specified in the prompt. Students need to paste the text they want to extract into the section labeled "Paste your text here".

5.3.3 Computer-aided translation

During the preview and review stages of interpreting training, learners need to prepare related parallel bilingual texts and build professional knowledge databases, which are crucial for both present and future use. CAT tools are essential to quickly align bilingual texts from interpreting textbooks to classroom teaching materials.

CAT tools can be categorized into multifunctional and single-function ones. Multifunctional tools not only include basic alignment and translation memory-building functions but also assist in translation, such as Trados, memoQ and Déjà Vu. Single-function tools, like TMXmall[6] (an online tool), ABBYY Aligner and Heartsome TMXeditor, focus on sentence alignment which can help learner to organize parallel sentence pairs.

Multifunctional and single-function software tools are powerful but require users to understand basic operating procedures and practice, especially for multifunctional software. With the assistance of large language models, this process becomes simplified. When both source and target language texts are available, learners can align sentence pairs when extracting terms by simply adding one to two conditions or keywords in the prompts. In cases where only the source language is available, large language models can help to translate into the corresponding target language and display in two columns when its prompts define language-specific domains and stylistic features. Xu et al. (2024) also provide an example as depicted in Prompt 4.

5.3.3.1   Prompt 4[7]

"User: Using the provided *English and Chinese* parallel texts, could you identify the *technical terms* related to *their general topic*? Please list these terms in a table with two columns: one for *English* terms and the other for their *Chinese* equivalents. If a term appears in only one language, translate it into the other language and include both the original term and its translation in the table." **"Paste your text here"**

Large language models can also be used for text alignment (which is a necessary step for building translation memory), allowing output in formats such as Excel or .tmx files as needed. However, large language models, such as GPT-4o, have relatively high requirements for the correspondence between the source text and the target text, as they are difficult to achieve alignment based on content recognition. However, the online alignment tool Tmxmall or alignment in Trados integrates an automatic alignment mechanism, where signal words like numbers and names can aid in identifying aligned content. Therefore, it seems CAT tools are relatively more available than large language models in this aspect.

5.3.4   Interpreting devices

In courses, such as simultaneous interpreting, over-the-phone interpreting (OPI) and video remote interpreting (VRI), learners need to practice using the corresponding technologies. These include simultaneous interpreting equipment, OPI systems and software platforms for VRI.

In simultaneous interpreting classes, various roles require the use of different tools. The speaker's voice (audio video record, or live conference) is transmitted through a microphone to the interpreter's workstation (students). The interpreters (students) then interpret in real time and send their voices through their own microphones. This interpretation is converted into a radio signal and transmitted via a transmitter. The audience (other students and teachers) uses receivers and headphones to receive and listen to both. This process requires highly synchronized and precise audio processing technology to ensure the accuracy of the interpretation. Commonly used simultaneous interpreting equipment includes Bosch,[8] GONSIN[9] and NewClass.[10] The operation of these systems is not complex; student interpreters still need to understand the basics and master the operations.

OPI is a remote interpreting service where an interpreter facilitates communication between parties speaking different languages via telephone. In OPI class, a three-way call is established between two clients (students) who need interpretation, and the interpreters (students). They may be required to know how to use the devices, and the apps, besides the liaison interpreting service.

VRI involves interpreters providing services from a distance using video remote communication technology. This is commonly used for meetings, seminars, court proceedings and medical consultations. In teaching, various platforms like Zoom, Boostlingo[11] and WebEx,[12] which support multiple languages, are used to create virtual classrooms, set course themes, facilitate group collaboration and discussion and record sessions.

With these changes in teaching content of interpreting, educators need to be proficient in the aforementioned technologies and integrate them effectively into the interpreting curriculum. Learners should also grasp the relevant technologies and principles. Teaching and learning complement each other.

It is worth noting that most of the above technology content learned in classroom is publicly accessible. Some instructors might use recordings, videos and textual materials from their own interpreting experiences as teaching cases. If these materials are still under privacy protection, it is crucial to be aware that online technologies are strictly regulated, especially when confidentiality agreements are involved. In actual interpreting work, if the client's texts are not allowed to be publicly available online or the papers have not yet been published, it is not permissible to use online terminology management tools or large language models, and the full text should not be directly uploaded. In such cases, offline terminology, alignment and CAT tools can be used, as online tools are subject to privacy protection restrictions.

5.4 Integration of technology and interpreting teaching models

In interpreting education, technology participates in classroom teaching activities, not only integrating into the teaching content but also bringing significant changes to teaching models and assessment methods. Interpreting courses, which are highly skill-based and training-intensive, emphasize a student-centered approach. To meet the needs of interpreting skills and thematic content teaching, AI involvement has introduced new changes to course teaching models and assessment methods.

5.4.1 Online courses

The use of flipped classrooms and blended learning models can strengthen the connection between in-class and out-of-class interpreting activities, combining classroom learning with skill training. For example, in a blended learning model, educators have students watch relevant course videos on Coursera, MOOC and SPOC platforms before teaching (see Figures 5.1 and 5.2). This allows students to engage in self-directed learning, understand the interpreting profession, background knowledge and basic interpreting skills and even perform some simple training exercises. These courses come with corresponding assessments, enhancing the efficiency and effectiveness of classroom teaching.

5.4.2 Virtual reality labs

Utilizing virtual reality (VR) in interpreting classroom settings enhances the authenticity, specificity and skill-directed focus of consecutive and simultaneous interpretation training. VR is situational, interactive, immersive and creative, and these diverse features allow students to engage in role-playing training, thereby facilitating interpretation practice. Liu (2018) and Deng and Liu (2020) analyzed the European Union (EU)-funded IVY training system (Tymczyoska et al., 2013), highlighting its capability for autonomous and collaborative learning among learners. According to Braun et al. (2014), IVY improves fundamental interpretation skills such as memory training and note-taking.

University of Geneva

International Organizations
for Interpreters

☆ **4.7** (89 reviews)

Beginner · Course · 1 - 3 Months

Peking University

计算机辅助翻译原理与实践
Principles and Practice of
Computer-Aided Translation

☆ **4.6** (77 reviews)

Intermediate · Course · 3 - 6
Months

FIGURE 5.1 Interpreting-related online courses on the platform of Coursera.[17]

FIGURE 5.2 Interpreting-related online courses on the platform of China University MOOC.[18]

The iLAB-X course platform integrates various teaching and training content and resources based on course themes and objectives (see Figure 5.3). It offers multiple immersive virtual simulation interpretation training courses, including the Immersive Virtual Simulation Scenario-based Interpretation and Intelligent Evaluation Project by the University of Electronic Science and Technology (2019),[13] the "Belt and Road" Business Interpretation Virtual Simulation Immersive Scenario Experimental teaching at Wenzhou University (2021)[14] and the English Contextual Emergency Interpretation Service Virtual Simulation Experiment at Shaanxi Normal University (2024).[15] These courses efficiently simulate scenarios using VR and AI technologies, some incorporating unexpected situations to enhance overall engagement. They comprehensively assess students'

FIGURE 5.3 Virtual reality (VR) learning platform.

interpretation accuracy, fluency, timeliness and professional ability to handle unforeseen circumstances.

These platforms not only offer diverse experimental modes but also enable teachers to tailor course content based on specific course characteristics, leveraging strengths from various sources for effective recombination. Furthermore, these platforms integrate technologies such as speech recognition and machine translation, providing technical support for terminology and digital aids during training. Their preliminary assessments and grading also provide evaluations based on student performance.

5.4.3 Digital resources for interpreting teaching

To meet the demand for realistic scenarios in interpreting training, VR presents one method with significant challenges due to its high cost and equipment requirements. A more feasible approach is the interpreting corpora and teaching resource repositories.

International organizations, such as the EU, place considerable emphasis on interpreting training. Since 2004, the EU has established the EU Speech Repository, which offers a multilingual and multi-themed corpus for consecutive and simultaneous interpretation practice. This repository categorizes materials into four difficulty levels: "Beginner", "Intermediate", "Advanced/Test-type" and "Very advanced" with both real-life speeches and pedagogical materials. Additionally, the EU has created the Online Resources for Conference Interpreter Training (ORCIT),[16] developed by the Directorate-General for Interpretation (DG Interpretation) of the European Commission. ORCIT provides high-quality teaching materials, methods and assessment tools tailored for interpreting teachers and trainers.

5.4.4 Generated teaching texts and audio-visual materials for interpreting

Acquiring authentic language data and audio-visual teaching cases requires educators to accumulate extensive resources or have access to live language data, along with proficiency in audio-visual processing technologies to ensure usability. With the advancement of large language models, their capabilities in processing and generating audio-visual content continue to strengthen, presenting opportunities for interpreting teaching.

The audio and video capabilities of large language models are highly robust. Xu et al. (2024) utilized ChatGPT Voice to set and test instructions capable of generating replicable audio for interpreting. Similarly, the video generation function of large language models is also powerful. Educators prepare video scripts prior to generation, outlining the curriculum topics and specific training skills with materials such as interpreting texts and audio. They then format the prepared materials and specific requirements into prompts for video AI, enabling the creation of pre-class or post-class extension training materials required by students. Furthermore, large language models can preliminarily assess students' interpreting assignments.

5.5 Insights toward the future interpreting education

In the future of interpreting education, technology will be deeply integrated with a focus on AI. Teachers will utilize a variety of AI technologies such as speech recognition, machine translation, VR and others to reshape classroom experiences, enrich course content and enhance teaching efficiency and quality. Students, in turn, will apply these technologies to improve their interpreting skills, expand their knowledge domains and cultivate essential professional competencies required of interpreters.

Courses will integrate AI-related content extensively, breaking traditional learning boundaries to support personalized learning paths and customized content, thereby providing students with a more tailored learning experience. Optimization of assessment methods will leverage advanced technologies and collaborative construction with interpreting teaching resources, advancing the development of automated scoring systems for interpreting. This will provide students with immediate and personalized assessment results, promoting comprehensive improvement in interpreting skills and fostering their application abilities.

5.6 Conclusion

This study discusses the influences toward interpreting education caused by technology which can be divided from two perspectives, functioning as the teaching contents and teaching methods. Technology as interpreting teaching contents plays a significant role in cultivating information technology literacy and interpreter professionalism. Technology as interpreting teaching methods contributes to an increase of student classroom engagement, effective skill training, personalized learning and multidimensional assessment.

Notes

1 All prompt examples are derived from Xu et al. (2024), who tested them through the ChatGPT 4.0. https://promptbank.unipus.cn/portal/detail?id=111.
2 https://promptbank.unipus.cn/portal/detail?id=111.
3 https://www.interpretbank.com/site/.
4 http://termbox.lingosail.com/.
5 https://promptbank.unipus.cn/portal/detail?id=112.
6 https://www.tmxmall.com/.
7 https://promptbank.unipus.cn/portal/detail?id=112.
8 https://www.bosch.com/.
9 https://www.gonsin.com/products/simultaneous-interpretation-system/.
10 https://www.newclass.com/dd.aspx?nid=3&typeid=12.
11 https://boostlingo.com/.
12 https://www.cisco.com/c/en/us/products/conferencing/webex-meetings/index.html.
13 https://www.ilab-x.com/details/page?id=3709&isView=true.
14 https://www.ilab-x.com/details/page?id=5712&isView=true.
15 https://www.ilab-x.com/details/page?id=12364.
16 https://orcit.eu/.
17 https://www.coursera.org/.
18 https://www.icourse163.org/.

References

Braun, Sabine, Catherine Slater, and Nicholas Botfield. 2014. "Evaluating the Pedagogical Affordances of a Bespoke 3D Virtual Learning Environment for Interpreters and Their Clients." In *Interpreter Education in the Digital Age*, edited by Suzanne Ehrlich and Jemina Napier, 39–67. Washington: Gallaudet University Press.

Deng, Juntao邓军涛, and Liu Menglian 刘梦莲. 2020. "Mianxiang kouyi jiaoxue de shipin yuliao ziyuanku shendu kaifa jizhi yanjiu." 面向口译教学的视频语料资源库深度开发机制研究 [Deep Processing Mechanisms of the Conversion of Video Corpora to Speech Repository for Interpreting Teaching]. *Waiyu jiaoyu yanjiu qianyan*外语教育研究前沿 [*Foreign Language Education in China*] 3(1): 37–43, 88.

Feng, Zhiwei 冯志伟, and Zhang Dengke 张灯柯. 2024. "Rengong zhineng zhong de da yuyan moxing." 人工智能中的大语言模型 [Large Language Model in Artificial Intelligence]. *Waiguo yuwen* 外国语文 [*Foreign Languages and Literature*] 40(3): 1–29.

Liu, Menglian 刘梦莲. 2018. "IVY xuni xianshi kouyi xunlian moshi yanjiu." IVY虚拟现实口译训练模式研究 [On the IVY Virtual Reality Interpreting Training Model]. *Shanghai fanyi*上海翻译 [*Shanghai Journal of Translators*] 5: 78–83.

Tymczyoska, Maria, Marta Kajzer-Wietrzny, Sabine Braun, Catherine Slater, Nicholas Botfield, and Margaret Rogers. 2013. Ivy–interpreting in virtual reality. Accessed May 20, 2025. https://www.researchgate.net/publication/309736364_IVY_Project_Interpreting_in_Virtual_Reality_-_Pedagogical_Evaluation_Report

Wang, Huashu, and Zhi Li. 2022. "Constructing a Competence Framework for Interpreting Technologies, and Related Educational Insights: An Empirical Study." *The Interpreter and Translator Trainer* 16(3): 367–390. https://doi.org/10.1080/1750399X.2022.2101850.

Xu, Jiajin 许家金, Zhao Chong 赵冲, and Sun Mingchen 孙铭辰. 2024. "Da yuyan moxing de waiyu jiaoxue yu yanjiu yingyong." 大语言模型的外语教学与研究应用 [Applications of Large Language Models in Foreign Language Teaching and Research]. Beijing 北京: Waiyu jiaoxue yu yanjiu chubanshe外语教学与研究出版社 [Foreign Language Teaching and Research Press].

6

EVOLVING CURRICULUM MODELS IN THE AI ERA

David B. Sawyer

The impact of artificial intelligence (AI) on society is profound; the media report on it daily. How profound is it? In their US national bestseller, *The Age of AI and Our Human Future*, Henry Kissinger, Eric Schmidt (of Google) and Daniel Huttenlocher (2021) state that "AI promises to transform all realms of human experience. And the core of its transformations will ultimately occur at the philosophical level, transforming how humans understand reality and our role within it" (p. 17). They describe the emergence of AI as bringing about a fundamental shift of our "sense of place in this world" (p. 194), upending how humankind has seen itself since the European Enlightenment. The authors state that

> [t]hroughout three centuries of discovery and exploration, humans have interpreted the world as Kant predicted they would according to the structure of their own minds. But as humans began to approach the limits of their cognitive capacity, they became willing to enlist machines – computers – to augment their thinking in order to transcend those limitations … as we are growing increasingly dependent on digital augmentation, we are entering a new epoch in which the reasoning human mind is yielding its pride of place as the sole discoverer, knower, and cataloger of the world's phenomena.
>
> *(pp. 49–50)*

The implications for the world of work are vast. In her article on "Machine learning, explained", Sara Brown (2021) from MIT's Sloan School of Management writes that machine learning (ML), a subfield of AI, "is changing every industry", and

> everyone in business is likely to encounter it and will need some working knowledge about this field. A 2020 Deloitte survey found that 67% of

DOI: 10.4324/9781003597711-7

companies are using machine learning, and 97% are using or planning to use it in the next year.

(n.p.)

According to Brown, researchers from the MIT Initiative on the Digital Economy "found that no occupation will be untouched by machine learning, but no occupation is likely to be completely taken over by it" (n.p.). This statement applies equally to translation and interpreting (T&I).

At the same time, language mediation services are and will continue to be a global growth industry, reaching almost 50 billion US dollars in 2022, according to Statista Research Department (2023), a leading provider of market and consumer data. According to their reporting, the market doubled in size over a ten-year period beginning in 2009 and suffered only slightly during the COVID-19 pandemic. The US Bureau of Labor Statistics (2024) estimates the employment of interpreters and translators in the United States to grow 2% from 2023 to 2033, slower than the average for all occupations. In its 2023 working paper "Not lost in translation: the implications of machine translation technologies for language professionals and for broader society", the OECD also projects stability and modest growth in the language professional occupations for other countries and regions around the world (Borgonovi et al. 2023). Importantly and perhaps counterintuitively from the lay perspective,

[a]nalyses based on data on online job vacancies for language professionals for a selected number of OECD countries between 2014 and 2019 indicate that the introduction of higher quality machine translation system did not lead to decreases in the demand for language professionals.

(p. 54)

As T&I continues to be a growth industry with job creation, key considerations for T&I educators are how the work will be done, the skills or competences required and perceptions about the role humans will have. At the 2023 annual conference of the American Machine Translation Association in Orlando, Translated (n.d.), a language services provider, put forward a metric called "Time to edit" (n.p.), the first estimate of the pace of progress toward what has been called the singularity in translation (see Melby and Hague 2019): the point at which the effort required to edit a machine translation (MT) would be the same as for a human translation done by a top professional. According to Translated's "Time to edit" metric, MT will be as good as human translation within the next ten years. The impact is that much larger volumes of text can be translated within the same budget, which can also benefit professional translators. Based on this trend, Translated estimates that there will soon be at least a tenfold increase in requests for professional translations and at least 100 times higher demand for MT.

However, text genre, complexity and use are key variables in any such analysis. In his seminal textbook on *Neural Machine Translation*, Koehn (2020) discusses the uses of MT and states that its "quality is not high enough that customers would pay a large amount of money for it. High-quality translation requires professional translators who are native speakers of the target language and ideally also experts in the subject matter" (pp. 20–21). While this statement may be self-evident for T&I educators, it is an important one to repeat when made by a leading researcher in the field of MT in a text intended for the next generation of neural machine translation (NMT) developers.

In *The Age of AI and Our Human Future*, Kissinger, Schmidt and Huttenlocher (2021) reference translation as a touchstone for the development of AI and ML applications. They write …

> Predicting the rate of AI's advance will be difficult. In 1965, engineer Gordon Moore predicted computing power would double every two years – a forecast that has proved remarkably durable. But AI progresses far less predictably. Language-translation AI stagnated for decades, then, through a confluence of techniques and computing power, advanced at a breakneck pace. In just a few years, humans developed AIs with roughly the translation capacity of a bilingual human. How long it will take AI to achieve the qualities of a gifted professional translator – if it ever does – cannot be predicted with precision.
>
> *(p. 86)*

Overall, AI/ML is driving growth, rather than inhibiting it, and transforming the language mediation professions. These developments make the emergence of AI/ML tools a watershed moment for T&I education. A closer look at the milestones of T&I education since the early 20th century places this transition point in its historical context.

When ask "What will be the impact of AI on the translation and interpreting professions?", what does ChatGPT say?

ChatGPT highlights several key factors as obvious advantages, including improved efficiency and accuracy, with much higher translation speed for large volumes of text, and ease of maintaining consistency of terminology and phrasing throughout a document or project, which can be challenging for human translators. AI will lead to changes in job requirements, with the role of translators and interpreters likely evolving to incorporate more knowledge of AI digital technology and data analysis. AI will create new job opportunities in MT quality assurance, post-editing and data training. Translators and interpreters will need to keep up with technological advancements to avoid being at risk of job displacement because it is impacting how services are provided.

ChatGPT is frank in its assessment of the limitations of AI/ML, including its struggles with the nuances of language and cultural context, concluding that AI technology will likely not completely replace human translators and interpreters. ChatGPT notes that it is important to recognize the value of human expertise in T&I, particularly in areas that require specialized knowledge and cultural context. Human expertise and creativity will still be needed for many tasks.

(Author's adaptation of ChatGPT 3.5 output)

6.1 The trajectory of T&I education: the AI milestone

Fundamental shifts in the historical trajectory of T&I education are typically driven by societal and technological forces external to the T&I professions, which precipitate field-internal developments. For example, the decline of French as a lingua franca in diplomacy and the rise of the conference industry in the late 19th and early 20th centuries led to more widespread use of language mediation at international events, resulting over time in the professionalization of T&I and, eventually, inclusion in academia. Similarly, contemporary use of the phrase *a Nuremberg moment* harkens back to the transition point after the Second World War when the appearance of simultaneous interpreting at the Nuremberg trials, "one of the first major international media events" (Behr 2015, 288), drew attention across societies. This technology application – widely perceived as new although already over a decade old – highlighted the need for trained interpreters, which accelerated the founding of programs and eventually the inclusion of simultaneous interpreting in T&I curricula. Alluding to the significance of this turning point, the phrase has been used over the last decade to imply the scope of change in interpreting services due to the rise of distance interpreting (Constable 2015), which accelerated due to the COVID-19 pandemic.

Major milestones in T&I education stem, among other things, from increased demand for T&I services, shifts in the type of service requested, the role of communication and digital technologies in making the provision of services more efficient and educators reacting to these developments. Key milestones in Table 6.1 are external to the profession and socially and/or technologically driven in nature: the growing recognition of the T&I professions and improvements in providing services, whether through the development of simultaneous interpreting equipment or personal computers, the internet and software solutions streamlining, standardizing and automating stages of the translation process.

Educators reacted to the professionalization of T&I first by taking steps to formalize training in university programs beginning in the 1930s and in accelerating numbers in the 1940s, after T&I became increasingly recognized as an occupation in its own right and the use of simultaneous interpreting at the Nuremberg trials

TABLE 6.1 Milestones in T&I education

Milestones in T&I education	
Early 20th century	T&I recognized as a profession in its own right
1940s	Universities establish programs in accelerating numbers
1945	Simultaneous interpreting equipment used at the Nuremberg trials
Early 1950s	Simultaneous interpreting added to T&I curricula
1953	First professional associations established – AIIC and FIT
Early 1970s	Translation studies defined as an academic discipline
1970s	Personal computers enable development of CAT tools
1990s	Distance learning program introduced through the internet
1990s	Academic programs established to train T&I educators
Late 1990s	International organizations set up networks – EMCI, EMT, UN consortium, PAMCIT
Since 2016	NMT and AI/ML automate translation in widely used software applications

was widely discussed and introduced in international and regional organizations. In the 1950s, interpreters and translators ensured the representation of their professions through the founding of associations like the *International Association of Conference Interpreters (AIIC)* and the *International Federation of Translators (FIT)*, which helped to define professional standards and promote working conditions and remuneration. The 1970s saw translation studies defined as an academic discipline, and university programs began offering specialized training for T&I educators two decades later.

In the late 1990s, international and regional organizations with in-house language mediation services began establishing educational networks, including the European Masters in Conference Interpreting (EMCI), the European Master's in Translation (EMT), the United Nations Universities Outreach Program and MoU network and the Pan-American Masters Consortium on Conference Interpretation and Translation (PAMCIT). These networks promoted high levels of training, the sharing of knowledge and best practices and opportunities for professional development.

Until Google's introduction of NMT in 2016, translation technology developed largely on a track separate from T&I education, with MT and computer-assisted translation (CAT) forming two lines of development (Wang and Sawyer 2023; O'Hagan 2019), and the impact on T&I education was limited. The development of MT applications began in the late 1940s, when concurrently the use of simultaneous interpreting equipment was becoming more widespread. In 1949, Weaver's Memorandum (1949) laid the foundation for rule-based MT, where language was translated based on a set of pre-defined rules. The Georgetown-IBM experiment in 1954 was the first public demonstration of MT; this system was also rule-based and translated only simple sentences in Russian-English (Hutchins 2006). Over a decade later, the report of the Automatic Language Processing Advisory Committee

(ALPAC) highlighted the limitations of rule-based MT and called for more research (Pierce et al. 1969). Progress was made incrementally through the 1980s. These systems were able to translate more complex sentences, but still suffered from accuracy and fluency issues.

The situation changed with the advent of personal computers and the internet, significant milestones for T&I education that would eventually enable distance learning programs, making T&I education more accessible all over the world. These technologies facilitated the development of CAT tools, which transformed the way translators work. After several decades of stagnation, in the 1980s, the first prototype of a CAT system, called TSS, significantly improved the efficiency and consistency of translation using databases, which enabled translators to store and retrieve previously translated segments of text. This allowed for real-time collaboration between translators and editors and paved the way for modern CAT tools. Trados, in-house tools like Transit, and others were introduced in the 1990s, which included features like translation memory, terminology management and project management. These developments resulted in the incorporation of translation technology, often as a standalone module or course, in T&I curricula, eventually to include for example localization, terminology management and project management as standard course offerings (O'Hagan 2019).

In 1993, IBM's introduction of statistical MT (SMT) was another turning point. This approach used statistical models to learn from vast amounts of bilingual data, resulting in higher accuracy and fluency. In 2003, neural language models began improving the ability of MT systems to process contextual information in the source text, resulting in another significant improvement in translation quality. Finally, in 2016, Google introduced NMT, which combined neural networks with SMT, a leap forward toward achieving the fully automated production of accurate and fluent translations that have the potential to reshape formal T&I education.

With the advent of AI/ML, the stakes are incredibly high for T&I professionals and T&I education. Despite the global growth trends outlined above, the social perception of T&I as professions and perhaps even the survival of programs are at stake. The risks are exemplified by the fact that futurists such as Ray Kurzweil (2005, 288) have long cited translation as a yardstick application for AI, similar to Kissinger, Schmidt and Huttenlocher's (2021) statement cited previously. Comparisons of human and machine output began with the question that Alan Turing put forward in his thesis "Computing machinery and intelligence" in 1950 – "Can machines think?" (p. 433) – which resulted in what became known as the Turing Test – can a machine "pass" as a human? In 1980, John Searle developed the Turing Test further in his thought experiment known as the Chinese Room Argument, in which the ability of a human locked away in a room to develop an analytical method to translate between Chinese and English without "understanding" the meaning of Chinese characters serves as an analogy for the way in which machines process language. Turing's Chinese Room Argument illustrates the point that human T&I serve as benchmarks in social perceptions of the quality and usefulness of AI systems. The emergence of generative AI *seems* to put an artificial general intelligence, or AGI, understood to mean AI systems capable of completing any intellectual task humans are capable of, within reach.

Although the precise trajectory is yet uncertain, it is accelerating, and the societal-level impact of AI also means that the ensuing changes to T&I education are fundamentally different in scope than most of the previous milestones in Table 6.1, as well as the "turns" of T&I studies (Snell-Hornby 2006). The latter have been described as paradigm shifts, or reorientations, in the academic study of T&I as cognitive and sociocultural phenomena (Pöchhacker 2015). The "turns" have had limited direct impact on curricula, curriculum models and instruction. The opposite is perhaps the case. As the field of T&I has expanded to include forms of interpreting other than conference, with designations like legal, health and public service, T&I educators have incorporated into curricula a wider variety of settings and social situations in which language mediation occurs, supporting at the same time reorientations of the lines of academic inquiry.

AI is a watershed due to the evolutionary change it imposes on technology-driven professions, dictated by external forces over which practitioners and educators have little control. Coupled with other factors impacting language mediation practices, such as the rise of remote interpreting and remote instruction during the pandemic and use of English as a lingua franca, it is worth restating that the transitions underway are at least on a par with the development of communication technology solutions for simultaneous interpreting used at the Nuremberg trials, which marked a "turning point in the history of simultaneous interpreting" (Behr 2015, 288). Its impact reshaped professional practice, societal views of it, nascent curricula and teaching methodologies, and created the foundations for nearly eight decades of professional conference interpreting (CI). With generative AI bursting into the collective consciousness, T&I educators are at a similar juncture today.

In his textbook *Introducing Interpreting Studies*, Franz Pöchhacker (2022) presents a visual of the levels of modeling interpreting in theory and research, with the human mind's "black box" at the center and layers expanding outward to add cognitive processing, elements of text and discourse, interaction between individuals in social and institutional settings and the role of interpreting as a profession and phenomenon in society. Following Kissinger et al.'s (2021) logic, it stands to reason that AI will impact all these levels from the neural through the anthropological.

Against the backdrop of global growth, including new or reframed employment opportunities driven by AI/ML, program administrators and curriculum designers can ask themselves how programs should adapt to this future. To remain relevant, language mediators will work increasingly with AI/ML, as implied in Brown's (2021) analysis. How must curriculum models evolve for T&I education to remain maximally relevant and successful in preparing students and graduates for future work? At the core of this question is the concept of T&I competence and how it will shift due to the societal forces of AI.

6.2   T&I competence and AI/ML

> The way to unleash machine learning success [is] to reorganize jobs into dis-
> crete tasks, some of which can be done by machine learning, and others that
> require a human.
>
> *(Brown 2021, n.p.)*

Professional competences in T&I refer to the knowledge, skills and abilities (KSAs)
required to perform these tasks at a professional or para-professional level. Grbić
and Pöchhacker (2015) point out that the notion of competences "is used in a number of different disciplines, and it is difficult to find a precise or commonly agreed
definition" (p. 69). Generally, though, in T&I studies, it has been used in research
on cognitive processes, education and certification programs, for example, aptitude testing and pedagogy (Grbić and Pöchhacker 2015). T&I studies often explore
distinctions between experts and novices, for example, research grounded in the
cognitive psychology of expertise, to attempt to explain developmental aspects of
skill acquisition, including distinctions between declarative and procedural knowledge, or "knowing what" in contrast to "knowing how", which also come to light
in interrelated KSAs.

As Amparo Hurtado Albir (2010) states in the entry on translation competence
in the *Handbook of Translation Studies*,

> Most models describing translation competence are componential models
> focusing on the components (or subcompetences) that characterize translation
> competence in written translation: language knowledge; extralinguistic knowl-
> edge; transfer competence; documentation skills; strategic competence …, etc.
>
> *(p. 56)*

Technology and technology use have generally played a secondary, subordinate
role in these descriptive models. For example, Pym's (2003) minimalist definition
of translation competence reduces the translation process to two overarching abilities: (1) generation competence: "the ability to generate a series of more than one
viable target text … for a pertinent source text" and (2) selection competence: "the
ability to select only one viable TT (target text) from this series and with justified
confidence" (p. 489). Although there is now growing interest and discussion of
translation technology in T&I pedagogy, including in discussions of competence
models (O'Brien and Rodríquez Vázquez 2020), researchers have also identified
this area as lacking in research on T&I curricula, at least until recently (Kenny
2020; Sawyer et al. 2019).

More elaborate models feature use of translation technology, although not
necessarily prominently. For example, the PACTE group's (2003) Translation
Competence Model, a widely cited theoretical framework developed by a group
of researchers from the Universitat Autònoma de Barcelona, aims to describe

the competences necessary for professional translation, based on a comprehensive analysis of translation processes and the skills required for successful translation. They identify model components including (1) *bilingual sub-competence* (procedural), (2) *extra-linguistic sub-competence*, (3) *knowledge about translation sub-competence*, (4) *instrumental sub-competence*, (5) *strategic sub-competence* and (6) *psycho-physiological components*. Each sub-competence is broken down into multiple subcomponents, with technology use addressed in the *instrumental sub-competence* as knowledge related to the use of documentation sources and information and communication technologies. Similarly, in Göpferich's (2009) translation competence model, developed in a longitudinal project called TransComp, many of the same or similar components appear, but also human factors, such as translation norms, self-concept and professional ethos and motivation. A comparison of these two models makes readily apparent that there is no consensus on the existence of specific modular competences and sub-competences, as Kiraly and Hofmann (2016) have noted. Rather, such models are simplifications of complex phenomena and hold instrumental power through their descriptive ability. In an illustration of this point, Massey (2018) recognizes the impact of digital technology, in particular the profound changes generated through AI/ML in translation, and calls for a reweighting of competences in such frameworks at the conclusion of his analysis of various models.

Taking "into account the research outcomes on translation and translator competence reported by the translation studies research community and the changes that have affected the language services industry", the 2022 update of the 2009 and then 2017 EMT competence framework has as its stated goal the consolidation and enhancement of "the employability of graduates of master's degrees in translation throughout Europe" (p. 3). It describes technology as one of five overarching competence areas, alongside knowledge, skills and abilities in language and culture, translation, personal and interpersonal and service provision. While, as expected, MT and knowledge and skills related to "present and future translation technologies" is included in the Technology category, MT also appears explicitly in the Translation category, as the EMT framework "also acknowledges that machine translation (MT) represents a growing part of translation workflows, and that MT literacy and awareness of MT's possibilities and limitations is an integral part of professional translation competence" (p. 7). The updated EMT competence framework thus draws attention to the increasing role of digital technologies in language mediation, which O'Hagan terms the "disruptive entanglement of human and machine" and a *sine qua non* for employment in any commercial translation service (O'Hagan 2019).

Re-consideration of Pym's (2003) two overarching competences makes the scope of the AI/ML milestone readily apparent. In terms of generation competence, AI/ML is an augmentation of human translation ability, with AI systems capable of processing previously inconceivable volumes of text. Yet for the time being, the uniquely human ability to produce and evaluate high-quality translations that require cultural knowledge, creativity and nuance cannot be matched by machines (Borgonovi et al. 2023). There are aspects of human generation competence that remain important for such texts. Selection competence is key as well in evaluating translated texts, although increasingly MT quality assessment and quality estimation are being automated through metrics such as Bilingual Evaluation Understudy, or BLEU, scores, (Human) Translation Edit Rates and Time-to-Edit in translation quality assessment workflows for suitable texts (Koehn 2020; Wang and Sawyer 2023).

Where the translation competence equation shifts dramatically is in how humans interact with machines, which is resulting in a new or modified form of human–computer collaboration or interaction (Läubli and Green 2019): "the translator's role is evolving, with tasks moving from translation proper to translation-related linguistic tasks" (Wang and Sawyer 2023, 167). Although ML leads to the atomization or reorganization of tasks as they have been conventionally executed, which Brown (2021) identifies as a factor for success in ML applications in the quotation at the start of this section, such a reorganization can only be accomplished by human subject-matter experts who possess an integrated, whole systems view and knowledge of translations workflows and processes. From this perspective, the pervasive impact of AI/ML in the coming years is a compelling reason to think increasingly of translation ability as an integrated, emergent set of skills, without distinct separation between components, which have been delineated differently across models – "an all-encompassing translator meta-competence" (Kiraly and Hofmann 2016, 85).

In research on spoken and signed language mediation, Grbić and Pöchhacker (2015) note that, as opposed to the work on competence in translation studies, the exploration of the construct of interpreting competence has been more limited in scope in interpreting studies and researchers tend to "speak of abilities and skills, presumably as a result of influential cognitive approaches to the study of interpreting processes" (p. 70). However, several competence models have been developed, including Pöchhacker's (2000) analysis noting a widespread consensus that the subcomponents include language and cultural skills, translational skills and subject-matter knowledge.

Kaczmarek (2010) draws on intercultural competence models to propose a competence model for community interpreting. This model features elements that are uniquely human, such as motivation and situational judgments of appropriateness, effectiveness and the role of context including culture and purpose that current AI/ML systems are unable to tackle. Building on Kalina's (2002) interpreting process model, Alba-Mikasa (2013) identifies five subcomponents, with each broken down into subskills. They include pre-process, in-process, peri-process, post-process and para-process skills. A review of the descriptions of these subcomponents brings to light where AI is already having an impact: assignment preparation in the pre- and post-process skill sets and use of digital tools supporting the peri-process skills.

Hence the discussion of the "augmented interpreter" and "AI boothmate" in venues such as AIIC UK and Ireland's 2023 webinar series on AI and interpreting. In terms of digital technologies competence, these developments and the rapid rise of distance interpreting and expansion during the COVID-19 pandemic are requiring new and evolving skill sets enabling interpreters to manage tools and platforms in a variety of environments and settings characterized at this juncture by diversity of applications and lack of standardization (see Braun 2020). It remains to be seen to what extent and how quickly AI applications will be able to replace human interpreters in the foreseeable future. Nevertheless, the technology for supporting routine, unnuanced, transactional exchanges is maturing rapidly.

Given its broad, societal impact, AI/ML will reshape all areas of translation competence, and eventually interpreting competence as well. It appears to be the case that the accelerating pace of innovation in communication and digital technologies will require more and faster adaptation of language mediation professionals going forward, and adaptability and flexibility will be decisive. Modeling translatorial action in the AI-driven workflow should become a productive area of research, as the scope of T&I studies continues to widen through the exploration of relationships between translation and communication and digital technologies (O'Hagan 2019). The need could arise to reframe translation/interpreting competence as a unitary language mediation competence with core language transfer tasks supported by AI/ML.

In this process, the investigation of how such competences can be transposed into curricular aims, goals and teaching objectives at the program level will require substantial development work. The EMT (2022) framework remains intentionally silent on curriculum design and implementation. Focusing on outcomes, the framework recognizes that curricula will differ, and some programs will offer instruction for a broader range of competences than others, and some competences may be acquired through different levels and types of curricula. Indeed, the curriculum in any institutional setting is driven by "that institution's educational philosophy, which is governed by specific political, cultural, legislative, and market-specific constellations and traditions in its country and region of the world" (Sawyer 2015, 99).

Looking beyond the conventional role of the translator handling directly the language transfer process, Melby and Hague (2019) review developments in the use of MT and discuss its potential impact on program goals. To ensure that curricula remain relevant in the age of MT, they propose a shift to expand the traditional concept of the translator to include responsibilities and tasks of a language services advisor. They write that "translator trainers must prepare students to enter a profession in which machine translation plays an increasingly important role", arguing that "modern translation students must prepare themselves to also be language-services advisors (LSAs)" who "recommend translation solutions before a project reaches a project manager. … Students who master the LSA role will be successful regardless of the evolution of "neural" networks and artificial intelligence" (p. 206). They see this skill set as falling under the PACTE group's *strategic sub-competence* although, as outlined above, it is doubtful that the influence of AI/ML will be limited to one sub-competence.

Given the trajectory of technology-related T&I milestones and the pervasive impact of AI/ML in society in general, it is unlikely that T&I curricula will remain maximally relevant to professional T&I practice and language mediation in society in general without a sweeping curriculum overhaul. Therefore, a factor to consider is how these new, shifting or expanding competences intersect with traditional curriculum models. With a nod to the caveat regarding the generalizability of curricula across institutional settings and the limitations of descriptive models previously stated, a reexamination of two widespread curriculum models described by Arjona (1984) can illuminate how the role of translation technology in T&I education has grown over the decades and how AI/ML competence in T&I curriculum models could further evolve – a thought experiment in curriculum design.

6.3 AI/ML competence in T&I curriculum models

Arjona (1984) bases her discussion of T&I curricula on the model that Velleman (Baigorri-Jalón 2015) introduced at the University of Geneva in 1941, which includes five components: area studies, multi-disciplinary studies, applied language arts and linguistic studies, practicum courses and deontology. She sequences instruction in translation (written) vs. interpreting (spoken) at the macro-level and identifies five structural models in place at the time: linear, modified linear, Y- or forked-track, modified Y-track and parallel track. These models are open and generic in nature and show only the relationship between these two overarching language mediation modalities, with instruction commonly in separate classrooms

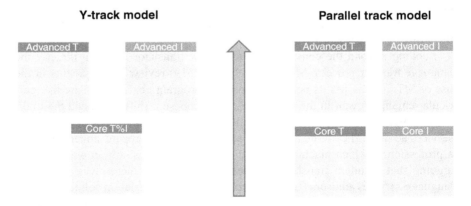

FIGURE 6.1 Arjona's Y-track and parallel track curriculum models.

for different cohorts, particularly at the advanced level. The models are above the level of courses and modules, with the latter often including several courses. In this sense, they show visually the overarching structure of a curriculum at the macro level and provide an alternative to depictions with a high degree of modularization, which Kiraly and Hofmann (2016) criticize as "rigid reglementation" that has been taken up in the "compartmentalization of knowledge" in some models of translation competence (p. 70). The following discussion uses the Y-track and the parallel track models as examples.

Arjona's original models in Figure 6.1 show a clear separation between instruction in translation (T) and in interpreting (I), with the separation at the beginning of instruction for the parallel track model and at the advanced level of instruction for the Y-track model. The implication is that there is limited intermingling between the two skill sets, particularly at the advanced level, in terms of competence, cohort and coursework.

When instruction in CAT tools was added to curricula – one of the milestones in the historical trajectory of T&I education – around the late 1980s into the 1990s, it tended to be offered to advanced students, usually in their second year of study of an M.A. program, or the equivalent (Figure 6.2). This changed, however, as the role of translation technology grew in the commercial sector and became more widely accepted and adopted by practitioners.

With the establishment of CAT and MT in professional practice, instruction in translation tools and related technology applications began to be offered earlier in the curriculum, including in standalone introductory courses, and programs incorporated instruction in project management and other translation technology courses, such as localization, at the advanced stages of the curriculum to facilitate greater specialization (Figure 6.3).

Subsequently, programs were augmented to offer tailored degrees with specializations in translation, localization and project management as program outcomes.

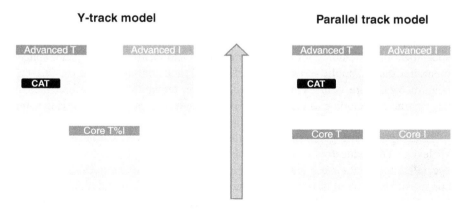

FIGURE 6.2 CAT in the Y-track and parallel track models.

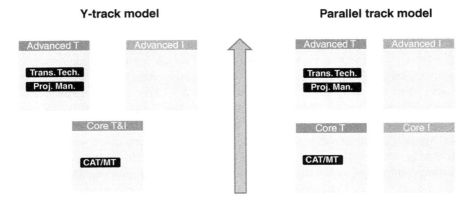

FIGURE 6.3 Translation technology in the Y-track and parallel track models.

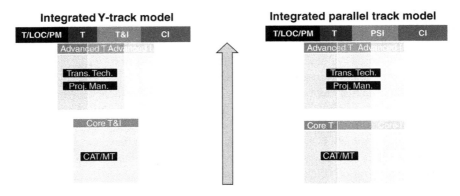

FIGURE 6.4 Translation technology in the integrated Y-track and parallel track models.

With increasing degree specialization and economies of scale in play for academic administrators, overlap of cohorts and curricula across degree tracks became more frequent, which can be reflected in integrated models showing four or more degree specializations, including for example translation and localization project management, translation, public service interpreting (PSI) and conference interpreting (CI) (Figure 6.4). Core courses related to translation technology are offered across degree tracks, including as electives for interpreting students. The result is in some cases an overlap throughout the curriculum, with shared core courses and electives.

Looking forward, given the wholesale impact that AI/ML will have on T&I at all levels, T&I educators can expect the pull toward an even greater technology specialization in their curricula. An adaptive response will be necessary to stay abreast of the general development of communication and digital technology use in society, which Kenny (2020) also identifies as a desirable curriculum outcome or competence.

Integrated Y-track model   **Integrated parallel track model**

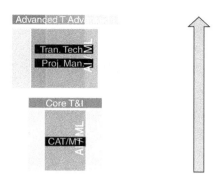 

FIGURE 6.5   AI/ML in the integrated Y-track and parallel track models.

Integrated Y-track model   **Integrated parallel track model**

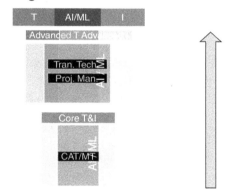 

FIGURE 6.6   AI/ML program outcomes.

In anticipation of these developments, there is a need for programs to infuse AI/ML in courses throughout the curriculum (Figure 6.5), even at the level of program outcomes. With the addition of a strong AI/ML component with stated outcomes at the program level, the Y-track and parallel track models evolve through further modifications, with AI/ML possibly substituting for language, business or policy studies in Arjona's original Velleman-based model. These summative AI/ML outcomes can be reflected in integrated curriculum models (Figure 6.6).

In addition to replacing some components in Arjona's original model, the heavier course load could be accommodated through joint degree or complementary certificate programs (Figure 6.7). Demand could also emerge for professional development offerings, as practitioners deepen and develop new AI/ML specializations. Beyond this stage of curricular evolution, the relative weight of T&I and AI/ML reverses in the models, with T&I becoming a component in AI/ML curricula.

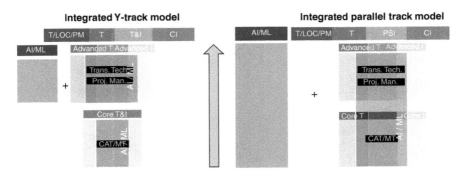

FIGURE 6.7 AI/ML as a standalone track.

6.4 Outlook

The shift to AI/ML and NMT applications impacts all practitioners, both new-comers and seasoned professionals. What will graduates be doing in the language industry and how can programs prepare them for new tasks and responsibilities? The shift brings with it a design challenge that can be met through needs and job analyses in the backward design approach widely recommended and followed in curriculum development (Wiggins and McTighe 2005). Job and market analyses can inform how skill sets are changing, what knowledge, skills and abilities should be emphasized, and the required level of attainment (Borgonovi et al. 2023; Saya-heen 2019).

The contours of several emerging developments are already clear: (1) T&I prac-titioners, and thus students, will benefit from a basic understanding of AI/ML oper-ations and workflow, which they increasingly should have opportunities to support and implement, and (2) AI/ML automation will relieve practitioners increasingly from the task of translation proper for many types of texts. This is already the case in many organizations where staff translators serve primarily as (senior) reviewers. Freed from the transfer step in the traditional translation workflow, practitioners should have the capacity to focus on refining competences to execute tasks where the performance of MT systems lags, such as refining finely nuanced translations using cultural expertise and creativity.

While learning to leverage AI/ML tools can augment the abilities of T&I prac-titioners, allowing them in turn to focus on the uniquely human capabilities that enable practitioners to do better than machines, the traditional roles of language services providers will expand, as new tasks and responsibilities are added (Melby and Hague 2019). Programs will thus have an opportunity to build on existing com-petences, such as continuous learning and cultural expertise, and introduce new ones related to AI/ML and natural language processing (NLP) (Wang and Sawyer 2023). To do this successfully, and help guide the processes that are redefining the T&I professions, educators will need to overcome barriers between themselves and AI developers working on NMT and ML solutions.

AI/ML developers and data scientists without formal translation or interpreting competence are often managing the development of these new translation systems and processes, as not many translators or interpreters have the deep expertise to do the development work on the machine side. However, as "creators, collectors, and managers of language data, translators and other domain experts have a significant role to play in machine training" (Wang and Sawyer 2023, 188). To achieve this goal, collaboration and communication are needed between computational linguists and translators as domain experts. This is a strong argument why T&I educators should take a hard look at curricula and consider how to integrate AI/ML content.

Whether the goal is to create a new program or revise an existing one, dialogue among stakeholders is essential. "[C]urricular frameworks emerge and evolve through a consensus-building process among all stakeholders, both internal and external to the educational institution" (Sawyer 2015, 96) and, for this reason, any discussion of curricular change, whether systemic or program specific, should encompass diverse voices that speak authoritatively to instructional requirements driven by the social and institutional context, as well as the social and market needs (Kelly 2010). This dialogue should now include AI and ML professionals.

Curriculum deliberations can also address how to ensure that the human value added is retained and augmented. As Kissinger et al. (2021) write, in AI/ML applications,

> Human developers will continue to play an important role in creation and operation. The algorithms, training data, and objectives for machine learning are determined by the people developing and training the AI, thus they reflect those people's values, motivations, goals, and judgment.
>
> *(p. 89)*

Machines do not possess values or judgment ability. However, AI/ML-driven translation systems should reflect the values and judgment of T&I specialists: the intentionality, creativity and human element of decision-making in language mediation processes. While the "entanglement of human and machine" is causing disruption (O'Hagan 2019) and may be generating concerns about the viability of language transfer skills-based T&I programs experiencing enrollment declines, language mediation educators have an opportunity to make their programs fit for the future by incorporating AI/ML components into their curricula.

References

AIIC UK and Ireland. 2023. *Artificial Intelligence and the Interpreter: AIIC UK and Ireland's Webinar Series on AI and Interpreting.* Accessed April 24, 2023. https://aiic.org/site/uk-ie/AI-interpreting

Alba-Mikasa, Michaela. 2013. "Developing and Cultivating Expert Interpreter Competence." *The Interpreters' Newsletter* 18: 17–34.

Arjona, Etilvia. 1984. "Testing and Evaluation." In *New Dialogues in Interpreter Education. Proceedings of the Fourth National Conference of Interpreter Trainers Convention*, edited by Marina L. McIntire, 111–38. Silver Spring: RID.

Baigorri-Jalón, Jesús. 2015. "Velleman." In *Routledge Encyclopedia of Interpreting Studies*, edited by Franz Pöchhacker, 432–33. London: Routledge.

Behr, Martina. 2015. "Nuremberg Trial." In *Routledge Encyclopedia of Interpreting Studies*, edited by Franz Pöchhacker, 288–89. London: Routledge.

Borgonovi, Francesca, Justine Hervé, and Helke Seitz. 2023. "Not Lost in Translation: The Implications of Machine Translation Technologies for Language Professionals and for Broader Society." *OECD Social, Employment, and Migration Working Papers* 291. OECD. https://doi.org/10.1787/e1d1d170-en.

Braun, Sabine. 2020. "Technology, Interpreting." In *Routledge Encyclopedia of Translation Studies*, edited by Mona Baker and Gabriela Saldanha, 569–74. London: Routledge.

Brown, Sara. 2021. "Machine Learning, Explained." MIT Sloan. Accessed April 10, 2023. https://mitsloan.mit.edu/ideas-made-to-matter/machine-learning-explained

Bureau of Labor Statistics, U.S. Department of Labor. 2024. *Occupational Outlook Handbook: Interpreters and Translators*. Accessed September 28, 2024. https://www.bls.gov/ooh/media-and-communication/interpreters-and-translators.htm.

Constable, Andrew. 2015. "Distance Interpreting: A Nuremberg Moment for Our Time." AIIC 2015 Assembly, Addis Ababa, January 18, 2015. https://aiic.org/document/9692/Constable_Distance%20Interpreting_A%20Nuremberg%20Moment%20for%20our%20Time_AIIC%202015%20Assembly_%20Day%203_Debate%20on%20Remote_01.18.2015.pdf

EMT. 2022. "European Masters in Translation Competence Framework 2022." Accessed April 24, 2023. https://commission.europa.eu/system/files/2022-11/emt_competence_fwk_2022_en.pdf

Göpferich, Susanne. 2009. "Towards a Model of Translation Competence and Its Acquisition: The Longitudinal Study TransComp." In *Behind the Mind: Methods, Models and Results in Translation Process Research*, edited by Susanne Göpferich, Amt Lykke Jakobsen, and Inger M. Mees, 11–38. Frederiksberg: Samfundslitteratur.

Grbić, Nadja, and Franz Pöchhacker. 2015. "Competence." In *Routledge Encyclopedia of Interpreting Studies*, edited by Franz Pöchhacker, 69–70. London: Routledge.

Hurtado Albir, Amparo. 2010. "Translation Competence." In *Handbook of Translation Studies (Vol. 1)*, edited by Yves Gambier and Luc van Doorslaer, 55–59. Amsterdam: John Benjamins.

Hutchins, John. 2006. "The First Public Demonstration of Machine Translation: The Georgetown-IBM System, 7th January 1954." Internet Archive Wayback Machine. https://web.archive.org/web/20160303171327/https://www.hutchinsweb.me.uk/GU-IBM-2005.pdf.

Kaczmarek, Lukasz. 2010. "Modelling Competence in Community Interpreting: Expectancies, Impressions and Implications for Accreditation." PhD dissertation, University of Manchester. https://pure.manchester.ac.uk/ws/portalfiles/portal/54593007/FULL_TEXT.PDF

Kalina, Silvia. 2002. "Quality in Interpreting and Its Prerequisites – A Framework for a Comprehensive View." In *Interpreting in the 21st Century: Challenges and Opportunities*, edited by Giuliana Garzone and Maurizio Viezzi, 121–30. Amsterdam: John Benjamins.

Kelly, Dorothy. 2010. "Curriculum." In *Handbook of Translation Studies (Vol. 1)*, edited by Yves Gambier and Luc van Doorslaer, 86–93. Amsterdam: John Benjamins.

Kenny, Dorothy. 2019. "Technology in Translator Training." In *The Routledge Handbook of Translation and Technology*, edited by Minako O'Hagan, 498–515. London: Routledge.

Kiraly, Don, and Sasha Hofmann. 2016. "Towards an Emergentist Curriculum Development Model for Translator Education." In *Towards Authentic Experiential Learning in Translator Education*, edited by Don Kiraly, 67–87. Mainz: Mainz University Press.

Kissinger, Henry, Eric Schmidt, and Daniel Huttenlocher, D. 2021. *The Age of AI and Our Human Future*. New York: Little, Brown and Company.

Koehn, Philipp. 2020. *Neural Machine Translation*. Cambridge: Cambridge University Press.

Kurzweil, Ray. 2005. *The Singularity Is Near: When Humans Transcend Biology*. London: Penguin.

Läubli, Samuel, and Spence Green. 2019. "Translation Technology Research and Human-computer Interaction (HCI)." In *The Routledge Handbook of Translation and Technology*, edited by Minako O'Hagan, 370–83. London: Routledge.

Massey, Gary. 2018. "Competences of T&I Graduates in the 21st Century." Lecture at Yangtze River Delta Forum: Educating "Languages+" Professionals for International Organizations, September 7–8, 2018. https://doi.org/10.13140/RG.2.2.11286.19521.

Melby, Alan K., and Daryl R. Hague. 2019. "A Singular(ity) Preoccupation: Helping Translations Students Become Language Services Advisors in the Age of Machine Translation." In *The Evolving Curriculum in Interpreter and Translator Education*, edited by David B. Sawyer, Frank Austermühl, and Vanessa Enríquez Raído, 205–28. Amsterdam: John Benjamins.

O'Brien, Sharon, and Silvia Rodríquez Vázquez. 2020. "Translation and Technology." In *The Routledge Handbook of Translation and Education*, edited by Sara Laviosa and Maria González-Davies, 264–77. London: Routledge.

O'Hagan, Minako. 2019. "Translation and Technology: Disruptive Entanglement of Human and Machine." In *The Routledge Handbook of Translation and Technology*, edited by Minako O'Hagan, 1–18. London: Routledge.

PACTE. 2003. "Building a Translation Competence Model." In *Triangulating Translation: Perspectives in Process-oriented Research*, edited by Fabio Alves, 43–66. Amsterdam: John Benjamins. https://www3.uji.es/~aferna/EA0921/3b-Translation-competence-model.pdf.

Pierce, John R., John B. Carroll, Eric P. Hamp, David G. Hays, Charles F. Hockett, Anthony G. Oettinger, and Alan Perlis. 1969. *Language and Machines: Computers in Translation and Linguistics.* ALPAC report, National Academy of Sciences, National Research Council, Washington, DC. https://nap.nationalacademies.org/resource/alpac_lm/ARC000005.pdf.

Pöchhacker, Franz. 2000. *Dolmetschen: Konzeptuelle Grundlagen und deskriptive Untersuchungen*. Tübingen: Stauffenburg.

Pöchhacker, Franz. 2015. "Paradigm." In *Routledge Encyclopedia of Interpreting Studies,* edited by Franz Pöchhacker, 293–95. London: Routledge.

Pöchhacker, Franz. 2022. *Introducing Interpreting Studies*, 3rd ed. London: Routledge.

Pym, Anthony. 2003. "Redefining Translation Competence in an Electronic Age: In Defense of a Minimalist Approach." *Meta* 48(4): 481–97. https://id.erudit.org/iderudit/008533ar.

Sawyer, David B. 2015. "Curriculum." In *Routledge Encyclopedia of Interpreting Studies,* edited by Franz Pöchhacker, 96–99. London: Routledge.

Sawyer, David B., Frank Austermühl, and Vanessa Enríquez Raído. 2019. "The Evolving Curriculum in Interpreter and Translator Education: A Bibliometric Analysis." In *The*

Evolving Curriculum in Interpreter and Translator Education, edited by David B. Sawyer, Frank Austermühl, and Vanessa Enríquez Raído, 1–22. Amsterdam: John Benjamins.

Sayaheen, Bilal. 2019. "Bridging the Gap Between Curricula and Industry: A Case Study of an Undergraduate Program in Jordan." In *The Evolving Curriculum in Interpreter and Translator Education*, edited by David B. Sawyer, Frank Austermühl, and Vanessa Enríquez Raído, 185–202. Amsterdam: John Benjamins.

Searle, John R. 1980. "Minds, Brains, and Programs." *Behavioral and Brain Sciences* 3(3): 417–24. https://doi.org/10.1017/S0140525X00005756.

Snell-Hornby, Mary. 2006. *The Turns of Translation Studies: New Paradigms or Shifting Viewpoints?* Amsterdam: John Benjamins.

Statista Research Department. 2023. "Market Size of the Global Language Services Industry from 2009 to 2019 with a Projection until 2022." *Statista*. Accessed April 24, 2023. https://www.statista.com/statistics/257656/size-of-the-global-language-services-market/.

Translated. n.d. *This Is the Speed at Which We Are Approaching Singularity in AI*. Accessed April 24, 2023. https://translated.com/speed-to-singularity.

Turing, Alan M. 1950. "Computing Machinery and Intelligence." *Mind* 59(236): 433–60. https://doi.org/10.1093/mind/LIX.236.433.

Wang, Peng, and David B. Sawyer. 2023. *Machine Learning in Translation*. London: Routledge. https://doi.org/10.4324/9781003321538.

Weaver, Warren. 1949. "Translation." Wayback Machine Internet Archive. https://web.archive.org/web/20120114183631/https://www.mt-archive.info/Weaver-1949.pdf.

Wiggins, Grant, and Jay McTighe. 2005. *Understanding by Design*, 2nd ed. Alexandria: Association for Supervision and Curriculum Development.

7

INTERPRETING ETHICS IN DIGITAL AGE

Considerations of using technology in interpreting profession

Shuxian Song and Heming Li

7.1 Introduction

Ethics refer to moral principles that control or influence a person's behavior, as defined in the *Oxford Advanced Learner's Dictionary* (9th Edition). These principles are not limited to personal conduct but extend to various professional domains. The concept of ethics has become a popular term in professional circles and academic fields for a few decades. Translation and interpreting (T&I) scholars and practitioners have eagerly embraced this idea to demonstrate a heightened understanding of their roles and obligations and have explored the ethical aspects of T&I from various perspectives.

The majority of research on ethical issues in interpreting is on the ethical obligations (or expectations) of interpreters, particularly in contexts where they have some discretion and autonomy, such as public or community interpreting (Barker 2005). This freedom allows them to make decisions independently when their moral principles are challenged (Angelelli 2004a, 2004b; Dean and Robert 2011; Inghilleri 2005, 2012; Ren 2010). The ongoing challenge lies in defining the interpreter's role in different public working conditions. This subject is also featured in major academic compilations such as the *Routledge Encyclopedia of Interpreting Studies* (Setton and Prunč 2016) and the *Handbook of Interpreting* (Ozolins 2015). The latter half of the 2000s witnessed the publication of volumes and monographs that treated ethics more thoroughly (e.g., Inghilleri 2012; Valero-Garcés 2014; Valero-Garcés and Tipton 2017). The release of the first monograph dedicated solely to ethics in public service interpreting (PSI) was in 2019 (Phelan et al. 2019), along with a special edition of *Translation and Interpreting Studies* on the ethics of non-professional T&I (Monzó-Nebot and Wallace 2020), highlighting continued scholarly interest in exploring this pivotal area.

DOI: 10.4324/9781003597711-8

Due to the constrained nature of their task and their limited agency, conference interpreters' ethical conduct has not been so extensively explored than that of public service interpreters. Apart from the established codes of ethics within T&I associations, there have been quite a number of educational and directive publications throughout the years addressing ethics in conference interpreting. *The Interpreter's Handbook: How to Become a Conference Interpreter* (Herbert 1952), represents one of the earliest comprehensive explorations into the requisite skills and responsibilities of a conference interpreter. Latter writings address various topics such as interpreter skill development (Seleskovitch 1998; Seleskovitch and Lederer 1995), professional loyalty, accuracy (Gile 1995), interpreter competence and qualifications (Kalina 2000) and the intersection of professionalism and ethics (Setton and Dawrant 2016a, 2016b). Some empirical studies explore instances where conference interpreters intentionally deviate from fidelity (Seeber and Zelger 2007) or neutrality (Zhan 2012; Wang and Feng 2017), not due to a lack of understanding or professional ethics, but rather as a conscious choice influenced by their cultural, national or institutional affiliations, providing a more nuanced understanding of how interpreters interpret these concepts in practice (Ren and Yin 2020).

Recently, more attention has been paid to the construction of interpreting ethics in the age of AI. Fantinuoli (2018) published *Interpreting and Technology,* exploring key issues, approaches and challenges in the interplay of interpreting and technology. *The Routledge Handbook of Translation and Ethics* (Koskinen and Pokorn 2021), offers a comprehensive overview of issues surrounding ethics in translating and interpreting. Horváth (2022) first examines how AI can be used in interpreting as well as the various tools already available, then discusses the ethical considerations raised by the use of AI in general and in interpreting in particular. *Interpreting Technologies - Current and Future Trends* by Pastor and Defrancq (2023) brings together a series of contributions on interpreting technologies focusing on interpreters' interaction with technological tools, interpreting services with technologies, as well as the relation between AI and interpreters. But exploring the ethics of interpreting in the digital age is still ongoing, and relevant ethical frameworks await to be reconstructed.

This chapter aims to amalgamate existing research on the ethical considerations in interpreting, identify challenges that interpreting ethics encounter in the context of modern interpreting technologies and encourage interpreting practitioners and researchers and affiliated organizations to contemplate and reconstitute the ethical standards of interpreting in the digital age by forecasting the trajectory of the evolution of interpreting ethics in the era of AI. After reviewing the development of ethical codes established for interpreters, three scenarios of technology utilization in the field of interpreting are introduced, including telephone interpreting (TI) and video remote interpreting (VRI), AI-enhanced computer-assisted interpreting (CAI) tools, as well as machine interpreting (MI). Notable ethical quandaries of interpreting in digital era are presented. Wide-ranging issues like faithfulness and neutrality, confidentiality and data security, intellectual property rights (IPR) protection, transparency and cultural awareness, combined with the discussion of the relationship between AI and humans will be involved. Alongside these challenges, we present strategic recommendations not only for interpreters but also for researchers and institutions involved in shaping the future of interpreting ethics.

7.2 Development of ethic codes in interpreting

Title 29 of Book Two of the *Leyes de las Indias*, a piece of legislation promulgated by the Spanish crown between 1529 and 1630, marks the origins of written guidelines, which regulate court interpreters' professional behavior in the Spanish colonies. Nonetheless, the guidelines mentioned earlier were established prior to the contemporary understanding of professional occupations, thus they cannot be classified as genuine ethical codes for professionals in their current context. During the mid-20th century, the T&I field began to gain legitimacy and sought to establish international standards, leading to the founding of the International Federation of Translators (FIT) in 1953. This marked the inception of the first global organization specifically dedicated to T&I professionals. That same year, the International Association of Conference Interpreters (AIIC) was founded, and the ethic codes of international conference interpreters were officially regulated by establishing the AIIC Code of Professional Ethics in early 1957. Other professional T&I organizations, both international and national, such as AUSIT (Australian Institute of Interpreters and Translators), American Translators Association, Association of Visual Language Interpreters of Canada and others, while not solely focused on conference interpreters, established their own ethical codes that govern the professional behavior of their members. The introduction of the Registry of Interpreters for the Deaf, Inc. (RID) Code of Ethics in 1965 represented a groundbreaking advancement in the professionalization of interpreting outside traditional international conference settings. While the initial focus was on ensuring the availability of competent interpreters, this Code played a crucial role in shaping the professional identity of sign language interpreters in North America. Over time, it has undergone revisions and updates, culminating in the development of the NAD (National Association of the Deaf)-RID Code of Professional Conduct.

The early 2000s marked a surge in interpreting ethics, with numerous documents emerging, such as APTIJ (Asociación Profesional de Traductores e Intérpretes Judicialesy Jurados) 2010, CHIA (California Healthcare Interpreting Association) 2002, NCIHC (National Council on Interpreting in Health Care) 2004, 2005, HIN (Healthcare Interpretation Network) 2007 and IMIA (International Medical Interpreters Association) 2007, and the following decade saw a second wave of revisions, with organizations updating their guidelines like AIIC 2015, 2018, AUSIT 2012 and NRPSI (National Register of Public Service Interpreters) 2016. The AIIC stands out as the sole global entity specifically focused on conference interpreters, emphasizing the ethical standards within this field. AIIC's Professional Standards (2015) address essential aspects such as contract agreements, remuneration, minimum team size and the number of required languages, ensuring high-quality service and optimal conditions for all conference stakeholders. By implementing uniform guidelines and granting a collective platform to both practitioners and those who utilize their services (AIIC 2015), AIIC aims not only to provide ethical guidance and elevate professional benchmarks for its members but also to exert regulatory influence over the entire field of conference interpreting. AIIC's Code of Ethics (2022) introduced rigorous standards of integrity, professionalism and confidentiality that all members must adhere to when performing conference interpreting. At the same time, standardization efforts for public service interpreting grew, including standards for community interpreting like ISO (International Organization for Standardization) 2014, interpreting services (ISO 2018) and legal interpreting (ISO 2019). Although the precise reasons for this convergence are speculative, it suggests a progression from theoretical principles (codes of ethics) to practical implementation (standards of practice) and eventually, industry-wide standards (Baixauli-Olmos 2020).

7.3 Use of technology in interpreting profession

The interpreting profession, a field historically characterized by its reliance on human skills of language proficiency and encyclopedic knowledge, has progressively embraced technological innovations. These advancements have not only facilitated greater efficiency but have also broadened access to interpreting services in unprecedented ways. As the profession continues to evolve, a deep understanding of current technological trends is essential for predicting future trajectories and preparing for emerging demands. Fantinuoli (2018) has predicted that technology-afforded interpreting may change both the interpreting ecosystem as well as its socio-economic aspects, in which three prominent technologies would build a huge impact on the trend of technological shift, i.e. remote interpreting (RI), CAI and MI.

7.3.1 Telephone interpreting and video remote interpreting

RI refers to the use of communication technologies to gain access to an interpreter in another room, building, town, city or country. In this setting, a telephone line or video conference link is used to connect the interpreter to the primary participants, who are together at one site (Braun 2015). TI and VRI are two popular methods of RI that provide language services that enable effective communication between individuals who are in different places. While both services aim to bridge language barriers and facilitate cross-cultural communication, they differ in the mode of interaction and level of visual cues available to interpreters and participants. TI is a form of interpreter-mediated, cross-cultural interaction, which takes place in conditions affected by the medium. It is a form of "technologized interaction" (Hutchby 2001). It allows individuals to connect with professional interpreters over the phone in real time. This RI solution is efficient, convenient and widely accessible, making it an essential tool for overcoming language barriers in various industries and settings. TI enables seamless communication in scenarios where visual cues are not crucial, such as in phone consultations, customer support and emergency situations. VRI involves using video conferencing technology to connect individuals with interpreters who provide real-time language support. It offers the added benefit of visual cues, facial expressions and body language, enhancing the accuracy and effectiveness of communication in situations where non-verbal cues are essential. With digitalization and extremely fast developing information and communication technologies, the availability of hardware and software enables dynamic monitoring and controlling of important parameters of VRI, such as lip synchronization, latency, video resolution and frequency response. Network infrastructures that allow for simultaneous transmission of high-definition video and high-quality audio signals via the Internet, combined with latest video, virtual reality and augmented reality technologies might offer possibilities to overcome existing technological, physiological and psychological problems.

In the past, VRI was gradually gaining traction within the conference interpreting landscape, albeit limited in scope compared to on-site interpreting assignments which were perceived as stable. However, the industry underwent a seismic shift with the onset of the COVID-19 pandemic (Cheung 2022). By the summer of 2020, VRI had rapidly surged to dominate the market, catching many conference interpreters off guard. This necessitated a rapid acquisition of new skills to remain competitive, as any delay in adapting meant potential income loss and diminished competitive edge. Moreover, VRI technology advanced rapidly, offering different systems such as all-in, side-by-side and add-in solutions, giving interpreters and clients a wide range of options.

This technological evolution introduced a multitude of features that had been initially overlooked, fundamentally reshaping the interpreting landscape. KUDO, Interprefy, Boostlingo, VerbalizeIt and Lingmo are the common cloud-based VRI platforms on the market today. Among those platforms, KUDO stands out for its user-friendly interface and scalability, while Interprefy is known for its seamless integration with video conferencing tools. Boostlingo provides on-demand interpreting services with integrated billing and scheduling features, making it popular in healthcare, legal and business settings. VerbalizeIt focuses on travelers and individuals, offering flexible on-demand interpreting via video or phone. Lingmo specializes in AI-powered real-time language translation services for businesses, delivering accurate and speedy communication solutions for global interactions.

Furthermore, the progression of VRI technology has catalyzed enhancements in other interpreting technologies, facilitating the seamless integration of multiple information streams and the provisioning of dedicated support tools like CAI tools for specialized tasks such as number and term recognition. Despite the oversight in readiness for the advent of VRI within the conference interpreting community, an opportunity remains to proactively prepare for the intricate technological milieu that will materialize through the integration of VRI and support tools.

7.3.2 AI-enhanced CAI tools

CAI tools are applications specifically designed to assist professional interpreters in at least one of the several sub-processes of interpreting, such as knowledge acquisition and management, lexicographical memorization, real-time terminology access and so on. Speech and text processing technologies have the potential to benefit both simultaneous and consecutive interpreting (Pöchhacker 2016). Currently, CAI tools are starting to offer advanced features based on recent developments in artificial intelligence, such as automatic speech recognition (ASR), machine translation and so on, sometimes becoming part of integrated platforms. Taking applications of AI on conference interpreting as examples, these technologies enable the production of "simultaneously interpreted" texts, which can be displayed on screens within conference venues, delivered through synthetic voices to the audience or both. The standard workflow encompasses several steps: AI interpreters initially decode speakers' messages using sophisticated speech recognition technologies; these decoded messages are then translated into target languages utilizing deep learning algorithms and extensive linguistic corpora; the translated texts are subsequently presented either on screens for attendees to read or broadcast via voice-over technologies for auditory consumption. The categorization of CAI tools and their expanded range as a part of ICT (Information and Communication Technology) is proposed to identify five main domains of ICT for interpreting (Guo et al. 2023). Despite the three domains of interpreting technology, i.e. CAI, RI and MI, which are identified by Fantinuoli (2018), this categorization adds computer-assisted interpreting training technologies and corpus-based interpreting technologies, which have gained traction over the last few years (Guo et al. 2023). CAI tools can be categorized into different generations (Fantinuoli and Montecchio 2021; Guo et al. 2023). The first-generation CAT tools are relatively conventional such as digital dictionaries, Word or Excel glossary lists and search engines. Second-generation CAI tools account for the immediacy of interpreting, which includes ASR software, note-taking applications like Notability, an application that only runs on IOS system, and some terminology management tools. As for the next generation of CAI tools, it is obvious that AI-enhanced CAI tools could accommodate more complex and context-based NLP features. By integrating new advancements in ML-based NLP, AI-enhanced tools aim at partially or fully automatizing some aspects of the interpreting workflow, from the preparation work to in-process and post-event activities.

Tablet interpreting is a subset of CAI. It refers to the practice of using a tablet device, such as an iPad or Android tablet, for interpreting services, which involves various interpreting modalities such as consecutive interpreting, simultaneous interpreting, Sim-Consec and other hybrid modes. Some research (e.g., Singer and Alexander 2016) has indicated variations in reading habits between traditional printed materials and digital media, but the convenience of accessing, annotating, storing and sharing digital content may present interpreters who are comfortable with technology with expanded opportunities (Drechsel 2019). Supporting tasks like accounting, production as well as interpreting teaching are also part of this field. However, tablet interpreting research is still in its infancy, and many promising research avenues remain, like analyzing how hybrid modalities like Sim-Consec and Sight-Consec affect notes, how tablets influence note-taking practices and delivery, as well as how to use an AI system on a tablet to assist interpreters. However, most tablet interpreting research so far has drawn on students working with the technology universities had available. Future research should focus on the uptake of professionals, considering the learning curve for practicing interpreters, seeing how easily professionals adapt to note-taking with a tablet, whether age, professional experience or novice-expert differences affect the ease and/or speed of tablet uptake and whether previous tablet use facilitates the adoption of tablet interpreting (Goldsmith 2023).

Another significant focus on the application of AI to interpreting is the automated evaluation of interpreter performance. Utilizing metrics like BLEU (bilingual evaluation understudy), METEOR (metric for evaluation of translation with explicit ordering), TER (translation error rate), or NIST (metric developed by National Institute of Standards and Technology), automated evaluation can aid trainers, recruiters or individuals in evaluating their own performance post-event. Han and Lu (2023) investigated the viability of using four automated machine translation evaluation metrics, i.e. BLEU, NIST, METEOR and TER, to assess human interpretation. They correlated the automated metric scores with the human-assigned scores (i.e. the criterion measure) from multiple assessment scenarios to examine the degree of machine-human parity and find that the reliability of these automated metrics relies heavily on their correlation with human assessments.

7.3.3 Machine interpreting

MI, also known as speech-to-speech translation, is an automated language translation process that converts spoken content from one language to another in the form of speech (Fantinuoli 2023). Unlike offline speech translation that caters to pre-recorded media, one of the key differentiation of MI lies in its real-time nature. It generates translations promptly during a live presentation, meant to be consumed immediately. The translation mode can either be consecutive, interpreting each sentence sequentially or simultaneous, enabling uninterrupted flow of speech throughout the process. For machine interpretation, there are two main strategies. The first, referred to as "end to end", uses a single component to transform spoken input straight into spoken output, eliminating the need for a text step in between. Despite being experimental and not widely used, it represents one extreme within the realm of MI. Conversely, the cascading approach, which is more commonly adopted, involves a modular process that usually includes steps like speech recognition, machine translation and text-to-speech synthesis (Sperber and Paulik 2020), forming the other end of the spectrum in MI techniques. And recent trends lie between these two extremes, aiming to merge some components of the cascading approach into single components. This reduces system complexity, enhances translation accuracy and naturalness and prevents error propagation between components (Gaido et al. 2020).

There are many MI tools that are widely used today. Google Translate, a ubiquitous choice, offers text, speech and image translation among over 100 languages, utilizing neural machine translation for enhanced accuracy. Microsoft Translator provides similar services with integration into various platforms, while DeepL Translator, known for its quality in European languages, employs a distinct neural network. Apps like iTranslate offer multiple modes, offline support and voice translation, making them convenient for travelers. Bing Translator, another Microsoft product, competes in the market. IBM's Watson Language Translator provides professional-grade translations with advanced NLP, and Pleco is a popular Chinese language learning app with an integrated translator.

More recently, at the launch of the generative language model ChatGPT 4o in May 2024, ChatGPT accomplished the task of simultaneous interpreting from English to Italian, demonstrating the power of its real-time interpreting. Its creator company, Open AI, has been focusing on improving the intelligence of its models over the past few years with an eye toward the future of human-computer interaction. The successful launch of ChatGPT 4o could herald a watershed moment that takes the paradigm of human-computer collaboration to a radical new stage and opens up limitless possibilities for the development of MI. In the past, several MI systems and AI models encountered numerous challenges. They often struggled with tasks like detecting tone, distinguishing between multiple speakers or handling background noise, and lacked the ability to produce laughter, music or convey emotions, resulting in subpar translations, significant delays and other issues. However, the advent of ChatGPT 4o marks a significant improvement, as it represents OpenAI's groundbreaking new model that integrates text, vision and audio processing seamlessly. This unified neural network architecture ensures that all inputs and outputs are managed consistently, thereby mitigating many of these previous limitations.

While MI tools have advanced significantly, they may not be suitable for every situation. Even with potential improvements in accuracy, tasks requiring deep understanding, emotional connections and personal responsibility still rely on human interpreters. As we move toward a future where humans and machines work together to access multilingual content, it's crucial to collaborate in guiding their use and establish ethical guidelines. Further research is needed to address questions such as how to improve AI for MI, the potential for using AI for simultaneous interpreting and how to teach MI in the age of AI.

7.4   Challenges of technology to interpreting ethics

The Internet accelerates the transmission of information but at the same time blurs the boundaries of personal power, the digital divide widens and people form social networks to communicate with others in the virtual world through technology that is different from interpersonal communication in reality, which leads to problems of identity, in particular the identity of machine translation or other technical tools for translation, unclear boundaries of power and responsibility, insufficient knowledge of people's own limitations and so on. The emergence of these problems indicates an imbalance between the development of technology and the development of human adaptability (Lu 2024). AI-powered interpreting encounters a spectrum of ethical challenges that necessitate careful consideration. These encompass issues such as ensuring accuracy and mitigating biases in interpretations, maintaining privacy and data security for users, striking a balance between AI and human oversight, promoting equitable access to interpreting services and fostering cultural sensitivity in language translations. Key ethical dilemmas include informed consent regarding AI interactions, the quality and reliability of AI interpretations across diverse language contexts, the potential impact on human interpreters' roles and livelihoods, the continuous monitoring and mitigation of biases in AI models, the responsiveness of AI in emergency situations and the recognition and respect of cultural nuances in interpretations. Addressing these ethical complexities demands a robust ethical framework, stakeholder collaboration and ongoing evaluation to uphold ethical standards in AI-powered interpreting.

7.4.1   *Faithfulness and neutrality*

In the digital age, faithfulness in interpreting activities becomes a multifaceted concern. First and foremost, the rapid advancements in AI have led to the development of translation tools that often lack the nuanced understanding and contextual adaptability found in human interpreters. These AI systems, while proficient in language processing, struggle with idiomatic expressions, colloquialisms and cultural references, which are essential for conveying meaning accurately. Secondly, the globalization of communication has increased the need for interpreters who possess cross-cultural knowledge and sensitivity. In the absence of a sufficient pool of culturally competent interpreters, digital platforms may inadvertently perpetuate stereotypes or misrepresent cultural practices, leading to misunderstandings or even offense. This calls for continuous training and development of AI models to incorporate diverse cultural perspectives. Moreover, the collaborative nature of interpreting, where humans and machines work together, can lead to inconsistencies if the human interpreter lacks proper guidance from AI or vice versa. Human interpreters need to be able to trust the technology while also exercising their judgment to correct errors or provide cultural contexts that the AI might miss.

To address these issues, the field of interpreting must adapt by integrating AI technologies with human expertise. This could involve developing hybrid systems that use AI for preliminary translations, followed by a human touch to refine and contextualize them. Additionally, investing in AI that learns from large datasets of multilingual and multicultural content, as well as incorporating user feedback, can improve its ability to adapt and become more faithful over time. Educational institutions and professional organizations should also prioritize the development of cultural competencies in AI developers and interpreters alike, fostering a shared understanding of the importance of cultural sensitivity in digital interpretation. This includes promoting interdisciplinary research and collaboration between linguists, technologists and cultural experts.

The issue of neutrality in interpreting activities in the digital age centers on the need for interpreters to be impartial and objective, ensuring that the information is unbiased and free from personal views, cultural biases or political influences. A major problem, as mentioned above, is algorithmic bias, which can creep into AI systems through biased training data. To address this issue, developers must employ diverse datasets and regular audits to minimize bias and maintain fairness. Human interpreters working with AI also play a crucial role in maintaining neutrality. They must be vigilant about introducing their own biases, adhere to strict guidelines and receive ongoing training to maintain cultural neutrality and avoid misunderstanding nuances. In addition to this, regulators and industry associations should collaborate to establish clear standards, guidelines and ethics for digital interpreting services.

7.4.2   Confidentiality and data security

Confidentiality has always been a major concern in ethical considerations of interpreting. This concern has become even more important with the rise of artificial intelligence, as it relates to the security of interpreting with technical support. The security concern encompasses not only the compromise of personal privacy for all parties involved in the interpreting process (such as interpreters, translation companies, clients and audience) but also the potential divulgence of sensitive information related to the activity. Take ASR function as example, which enables the live transcription of speeches in AI-based terminology software developed for conference interpreters working in the simultaneous mode. In InterpretBank's workflow, the interpreter's headset captures audio, which is then forwarded to the computer's sound card featuring the ASR-CAI software. This audio is processed by the InterpretBank API, hosted on servers in Dresden, Germany, providing an immediate transcript of the spoken words. InterpretBank favors the Google Cloud Speech-to-Text API2 for its ASR tasks (Defrancq and Fantinuoli 2020). In the context of data privacy and protection, the trustworthiness of the data transmission may be of concern. First, since the sound signal is transmitted from the interpreter's computer to the server, interpreters can easily and unintentionally violate the principle of confidentiality. Secondly, they could also inadvertently violate the rights of the speaker. It is paramount to guarantee that the interpreting tools utilized do not compromise information security and that the platforms used are free from bugs. Moreover, it is imperative to ascertain that information is securely stored in the cloud and adequately protected against cyber threats. Furthermore, the effective management of information usage is a matter of significant concern. Technological advancements pose significant challenges due to their involvement with managing vast, complex datasets, known as big data, that traditional data-processing applications cannot handle. Training future interpreters now must address not only the sheer scale of these data sets and the technological proficiency required, primarily for machine translation, but also the handling of potentially sensitive information within the realm of cybersecurity. The intricate task of accessing, transmitting or managing such data carries substantial legal consequences. Efforts to regulate these emerging issues will invariably affect all parties dealing with big data, including translators. Consequently, this will also introduce new challenges in the education of interpreters, concerning both the operation of cutting-edge technological resources and the ethical considerations involved.

7.4.3 IPR protection

The issue of IPR in the era of artificial intelligence is an emerging concern. Although there are currently no established norms for the protection of IPR in interpreting in the age of AI, several issues deserve careful attention. Yanisky-Ravid and Martens (2019) put forward an issue that some top translation companies such as Google Translate, Microsoft Translator, DeepL and Systran translate copyrighted written materials (e-mails, advertisements, news, articles, songs, literary works, books, etc.) in order to create derivative works under the current US copyright system. However, these companies typically do not contact rights holders for permission, which means that they are "neither licensed to use the works nor paid royalties for them" (Yanisky-Ravid and Martens 2019, 103). As a result, many companies violate copyright laws on a daily basis. By the same token, an interpreter should be careful about copyright before, during and after the interpreting process.

Firstly, interpreters must ensure they have the proper licenses or permissions to interpret copyrighted materials in digital form such as software interfaces, presentations or websites to avoid potential claims of copyright infringement. Secondly, in the case of machine-assisted interpreting, the ownership of the interpretation itself becomes an issue, and relevant guidelines have not been clearly defined. Additionally, recording interpretations for future use or training may also raise intellectual property concerns, particularly regarding obtaining explicit consent and awareness of recording policies. Moreover, interpreters working through digital platforms are often subject to terms of service, which may include IPR terms, and failing to review these agreements may lead to infringement and harm their interests.

7.4.4 Transparency

Another ethical concern related to AI-based tools in interpreting is transparency. Transparency of interpreting processes and decision-making in AI systems is one of the basic criteria for AI ethics. Bostrom (2011) proposed that there will be an increasing need to develop artificial intelligence models that not only have remarkable power and scalability but also are transparent, allowing them to be readily reviewed or audited. The essence of transparency in AI systems lies in their ability to provide a comprehensive understanding, by enabling users to scrutinize, replicate and trace the inner workings of decision-making processes and learning algorithms. It encompasses the revelation of the data's origins, evolution and how it influences the system's functioning. Additionally, transparency necessitates clear communication regarding the choices made regarding data sources, development procedures and ensuring active participation and collaboration from all stakeholders who contribute to the system's creation and operation. In essence, it's about fostering a sense of openness and accountability in the AI process (Dignum 2019, 54).

When combining with interpreting, transparency is essential for interpreter to build trust with AI, which means that "it should always be possible to find out why an autonomous system made a particular decision, especially if that decision caused harm" (Bird et al. 2020, 31). First, AI systems should be able to trace and explain the steps taken in interpreting data or perform specific tasks, which are all under the control of human interpreters and other related parties. It helps these users understand how and why AI systems make certain interpretations. This understanding fosters trust and allows users to verify the accuracy and fairness of the interpretations provided. It also enables users to identify any biases or errors in the AI system's decision-making and take appropriate actions. Second, transparency contributes to the development of robust AI systems. By making the decision-making process transparent, developers, researchers, as well as interpreters can identify and address any vulnerabilities or limitations in the AI models. This iterative process enhances the further development of AI systems. Besides that, AI systems that operate with transparency are more likely to comply with legal and ethical standards. It allows for better regulation and policy development to ensure that AI technologies are used responsibly and ethically.

7.4.5 Cultural awareness

Although the current AI is "omniscient", it does not fully understand the real cultural meaning behind certain languages, which can easily lead to ambiguity in interpreting activities. Citing the example of interpreting in crisis situations, misinterpretations due to the nuances of language and cultural context can lead to dangerous misunderstandings, inappropriate medical interventions and misallocated aid, thereby exacerbating the crises. Furthermore, the absence of cultural competence and empathy in automated systems can alienate affected individuals, while the sidelining of human interpreters diminishes the nuanced understanding and compassionate communication essential in such high-stakes environments. The challenge of attributing accountability in the event of errors also complicates the situation, as the inability to promptly correct mistakes can delay crucial interventions and undermine the credibility of response efforts. Thus, addressing these ethical concerns entails a judicious combination of technology and human insight to ensure effective, sensitive and accountable crisis response.

Beyond the existing limitations of AI in accommodating multilingual and multicultural interpretive contexts, a significant concern arises from the apparent deficit in cultural sensitivity. This shortcoming, prevalent across various interpreting modalities, including collaborative, machine and human interpreting, has the potential to give rise to misinterpretations, inappropriate utterances and, in extreme cases, grave repercussions. The primary issue concerns the phenomenon of stereotype and prejudice within AI-driven interpreting, due to its reliance on data processing and susceptibility to amplified biases from internet content. To mitigate this, interpreters must exhibit self-awareness of their cultural biases and diligently strive to minimize them during interpreting tasks, concurrently verifying the accuracy of AI-generated insights. The scarcity of online resources, particularly for niche cultures or minority ethnicities, poses a challenge, as it limits the AI's corpus and can inadvertently perpetuate stereotypes by reinforcing under-represented identities. Another critical aspect is digital etiquette, reflecting the varying cultural norms surrounding online communication, such as tone, response speed and formality. Interpreting in digital environments necessitates adaptability to these nuances, a skill that human interpreters excel in, particularly in real-time interpreting where they adeptly navigate cultural shifts like humor and linguistic taboos. AI currently lags behind in this regard, leaving substantial room for improvement. Moreover, cultural attitudes toward technology differ across nations and societies, necessitating interpreters' understanding of potential technological barriers or user preferences. This knowledge is crucial in ensuring seamless and culturally sensitive digital interactions. In light of these complexities, the enhancement of AI's cultural awareness transcends technological advancements, requiring ongoing and targeted professional development for interpreters.

As discussed above, primary ethical issues within the field of machine/AI interpreting are as follows: The accountability for inaccuracies in AI-generated interpretations presents one major challenge in the AI era. Unlike human interpreters, AI systems cannot be held accountable for errors, which can lead to serious misunderstandings, particularly in critical contexts such as diplomatic or legal settings. Issues related to confidentiality and IPR are prominent in both data collection and software design. The extensive data required to train AI systems often includes sensitive information, thereby raising concerns about data privacy and ownership. Digital and legal tools were also needed to balance competing interests, and policymakers should develop norms and guidelines. Given the global nature of the issue, it is hoped that relevant guidelines may be issued by the World Intellectual Property Organization (WIPO). Transparency in interpreting activity in the AI era is crucial for building trust, ensuring accountability, improving system reliability and adhering to ethical and regulatory standards. It promotes a better understanding of AI systems and helps to address concerns or challenges that arise from their use. Cultural awareness issues are also magnified in the digital environment. Failure to give high priority to this issue, whether by researchers, interpreters themselves or AI technology developers, can lead to failure in interpreting tasks. Deeper learning is needed on all sides to meet this challenge. Whether it is building a larger corpus or improving the accuracy of the application, it is an area worth exploring further.

In addition to the challenges mentioned above, another issue that has been plaguing the digital age is the ethical dilemma arising from the interaction between human interpreters and AI in the field of interpreting. The prevailing view is that AI and human interpreters should collaborate to enhance each other's capabilities rather than compete. However, practical implementation poses several challenges. There are concerns about job displacement, as organizations may prioritize cost-effective AI solutions over human expertise. The current limitations of AI in understanding context, cultural nuances and emotional tones necessitate human oversight to ensure accurate and appropriate interpretations. Additionally, effective collaboration requires substantial investment in training and development for both human interpreters and AI systems. Addressing all these ethical complexities demands a robust ethical framework, stakeholder collaboration and ongoing evaluation to uphold ethical standards in AI-powered interpreting.

7.5 Conclusion

Although interpreting researchers and organizations have studied in depth the issue of interpreting ethics from different perspectives and set up a series of norms and regulations, which continue to update until now. AI era signifies an impending revolution in these standards. This chapter points out what challenges interpreters may face under the impact of AI, some of which are newly born with the application of AI/machines in the field of interpreting, and others that have already existed such as fidelity and confidentiality are interpreted in a new way in the digital age. These issues have implications not only for individual interpreters but also for the interpreting profession as a whole. The key to overcome these ethical problems relies on figuring out the relationship between human and machine. How to make good use of AI/machines for interpreting activities and how to realize good human-machine coordination will be one of the significant directions for further research. Along with the innovation of AI, new ethical issues may come up accordingly, which means that there exists a dearth of systematic regulations of the ethical considerations associated with the integration of technology in interpreting activities, and feasible solutions to the ethical challenges posed by technological advancements.

Funding

The research is funded by the Humanities and Social Science Research Fund of the Ministry of Education of China (Grant No. 23YJC740054).

References

AIIC. 2015. "Professional Standards." Accessed December 26, 2024. https://aiic.africa/wp-content/uploads/2020/12/professional-standards-eng.pdf.

AIIC. 2022. "AIIC Code of Professional Ethics." Accessed December 26, 2024. https://aiic.org/document/10277/CODE_2022_E&F_final.pdf.

Angelelli, Claudia V. 2004a. *Revisiting the Interpreter's Role: A Study of Conference, Court, and Medical Interpreters in Canada, Mexico, and the United States.* Amsterdam: John Benjamins Publishing Company.

Angelelli, Claudia V. 2004b. *Medical Interpreting and Cross-cultural Communication.* Cambridge: Cambridge University Press.

Baixauli-Olmos, Lluís. 2020. "Ethics Codes for Interpreters and Translators." In *The Routledge Handbook of Translation and Ethics,* edited by Kaisa Koskinen and Nike K. Pokorn, 297–319. London and New York: Routledge.

Barker, Chris. 2005. *Cultural Studies: Theory and Practice.* London: Sage.

Bird, Eleanor, Jasmin Fox-Skelly, Nicola Jenner, Ruth Larbey, Emma Weitkamp, and Alan Winfield. 2020. "The Ethics of Artificial Intelligence: Issues and Initiatives." European Parliamentary Research Service. https://www.europarl.europa.eu/thinktank/en/document/EPRS_STU(2020)634452.

Bostrom, Nick. 2011. "Infinite Ethics." *Analysis and Metaphysics* 10: 9–59.

Braun, Sabine. 2015. "Remote Interpreting." In *The Routledge Handbook of Interpreting,* edited by Holly Mikkelson and Renée Jourdenais, 401–16. London and New York: Routledge.

Cheung, Andrew K. F. 2022. "COVID-19 and Interpreting." *INContext: Studies in Translation and Interculturalism* 2(2): 9–13.

Dean, Robyn K., and Robert Q. Pollard. 2011. "Context-Based Ethical Reasoning in Interpreting: A Demand Control Scheme Perspective." *The Interpreter and Translator Trainer* 1(1): 155–82.

Defrancq, Bart, and Claudio Fantinuoli. 2020. "Automatic Speech Recognition in the Booth: Assessment of System Performance, Interpreters' Performances and Interactions in the Context of Numbers." *Target* 33(1): 73–102.

Dignum, Virginia. 2019. *Responsible Artificial Intelligence: How to Develop and Use AI in a Responsible Way.* Berlin: Springer.

Drechsel, Alexander. 2019. "Technology Literacy for the Interpreter." In *The Evolving Curriculum in Interpreter and Translator Education: Stakeholder Perspectives and Voices,* edited by David B. Sawyer, Frank Austermühl, and Vanessa Enríquez Raído, 259–68. Amsterdam: John Benjamins Publishing Company.

Fantinuoli, Claudio. 2018. *Interpreting and Technology.* Berlin: Language Science Press.

Fantinuoli, Claudio. 2023. "Towards AI-Enhanced Computer-Assisted Interpreting." In *Interpreting Technologies – Current and Future Trends,* edited by Gloria Corpas Pastor and Bart Defrancq, 46–71. Amsterdam and Philadelphia: John Benjamins Publishing Company.

Fantinuoli, Claudio, and Maddalena Montecchio. 2021. "Defining Maximum Acceptable Latency of AI-Enhanced CAI Tools." Paper presented at TechLing 2021: 6th International Conference on Language, Linguistics, and Technology, Vigo, Spain, November 2021.

Gaido, Marco, Beatrice Savoldi, Luisa Bentivogli, Matteo Negri, and Marco Turchi. 2020. "Breeding Gender-Aware Direct Speech Translation Systems." Paper presented at the 28th International Conference on Computational Linguistics, Barcelona, Spain, December 2020. https://aclanthology.org/2020.coling-main.350.

Gile, Daniel. 1995. *Basic Concepts and Models for Interpreter and Translator Training.* Amsterdam and Philadelphia: John Benjamins Publishing Company.

Goldsmith, Joshua. 2023. "Tablet Interpreting: A Decade of Research and Practice." In *Interpreting Technologies – Current and Future Trends,* edited by Gloria Corpas Pastor and Bart Defrancq, 27–45. Amsterdam and Philadelphia: John Benjamins Publishing Company.

Guo, Meng, Lili Han, and Marta Teixeira Anacleto. 2023. "Computer-Assisted Interpreting Tools: Status Quo and Future Trends." *Theory and Practice in Language Studies* 13(1): 89–99.

Han, Chao, and Xiaobei Lu. 2023. "Can Automated Machine Translation Evaluation Metrics be Used to Assess Students' Interpretation in the Language Learning Classroom?" *Computer Assisted Language Learning* 36(5–6): 1064–87.

Herbert, Jean. 1952. *The Interpreter's Handbook: How to Become a Conference Interpreter,* 2nd ed. Genève: Editions Georg.

Horváth, Ildikó. 2022. "AI in Interpreting: Ethical Considerations." *Across Languages and Cultures* 23(1): 1–13.

Hutchby, Ian. 2001. *Conversation and Technology.* Cambridge: Polity.

Inghilleri, Moira. 2005. "Mediating Zones of Uncertainty: Interpreter Agency, the Interpreting Habitus and Political Asylum Adjudication." *The Translator* 11(1): 69–85.

Inghilleri, Moira. 2012. *Interpreting Justice: Ethics, Politics and Language.* London and New York: Routledge.

ISO [International Organization for Standardization]. 2018. *Interpreting Services: General Requirements and Recommendations.* www.iso.org/standard/63544.html.

ISO [International Organization for Standardization]. 2019. *Interpreting Services: Legal Interpreting: Requirements.* www.iso.org/standard/67327.html.

Kalina, Sylvia. 2000. "Interpreting Competences as a Basis and a Goal for Teaching." *The Interpreters' Newsletter* 10: 3–32.

Koskinen, Kaisa, and Nike K. Pokorn. 2021. *The Routledge Handbook of Translation and Ethics,* 1st ed. London and New York: Routledge.

Lu, Yan 陆艳. 2024. "Rengongzhineng shidai fanyi jishu lunli goujian" 人工智能时代翻译技术伦理构建 [Ethical Construction of Translation Technology in the Age of Artificial Intelligence]. Zhongguo fanyi 中国翻译 [*Chinese Translators Journal*] 45(1): 117–25.

Monzó-Nebot, Esther, and Melissa Wallace. 2020. "Ethics of Non-Professional Translation and Interpreting." *Special Issue of Translation and Interpreting Studies* 15(1): 1–14.

Ozolins, Uldis. 2015. "Ethics and the Role of the Interpreter." In *The Routledge Handbook of Interpreting,* edited by Holly Mikkelson and Renée Jourdenais, 319–36. London and New York: Routledge.

Pastor, Corpas G., and Bart Defrancq. 2023. *Interpreting Technologies – Current and Future Trends.* Amsterdam and Philadelphia: John Benjamins Publishing Company.

Phelan, Mary, Mette Rudvin, Hanne Skaasden, and Patrick S. Kermit. 2019. *Ethics in Public Service Interpreting.* Boca Raton: CRC Press.

Pöchhacker, Franz. 2016. *Introducing Interpreting Studies,* 2nd ed. London and New York: Routledge.

Ren, Wen. 任文. 2010. *Lianluo kouyi guocheng zhong yiyuan de zhutixing yishi yanjiu* 联络口译过程中译员的主体性意识研究 [The Liasion Interpreter's Subjectivity Consciousness]. Beijing 北京: Waiyu jiaoxue yu yanjiiu chuban she 外语教学与研究出版社 [Foreign Language Teaching and Research Press].

Ren, Wen, and Mingyue Yin. 2020. "Conference Interpreter Ethics." In *The Routledge Handbook of Translation and Ethics,* edited by Kaisa Koskinen and Nike K. Pokorn, 195–210. London and New York: Routledge.

Seeber, Kilian G., and Christian Zelger. 2007. "Betrayal-Vice or Virtue? An Ethical Perspective on Accuracy in Simultaneous Interpreting." *Meta* 52(2): 290–98.

Seleskovitch, Danica. 1998. *Interpreting for International Conferences.* Washington: Pen and Booth.

Seleskovitch, Danica, and Marianne Lederer. 1995. *A Systematic Approach to Teaching Interpretation.* Paris: The Registry of Interpreters for the Deaf.

Setton, Robin, and Andrew Dawrant. 2016a. *Conference Interpreting: A complete Course.* Amsterdam and Philadelphia: John Benjamins Publishing Company.

Setton, Robin, and Andrew Dawrant. 2016b. *Conference Interpreting: A Trainer's Guide.* Amsterdam and Philadelphia: John Benjamins Publishing Company.

Setton, Robin, and Erich Prunč. 2016. "Ethics." In *The Routledge Encyclopedia of Interpreting Studies,* edited by Franz Pöchhacker, 144–48. London and New York: Routledge.

Singer, Lauren M., and Patricia A. Alexander. 2016. "Reading Across Mediums: Effects of Reading Digital and Print Texts on Comprehension and Calibration." *The Journal of Experimental Education* 85(1): 155–72.

Sperber, Matthias, and Matthias Paulik. 2020. "Speech Translation and the End-to-end Promise: Taking Stock of Where We Are." Paper presented at the 58th annual meeting of the Association for Computational Linguistics (ACL), Seattle, WA, July 2020. https://aclanthology.org/2020.acl-main.661.

Valero-Garcés, Carmen. 2014. *(Re)visiting Ethics and Ideology in Situations of Conflict.* Alcalá: Universidad de Alcalá.

Valero-Garcés, Carmen, and Rebecca Tipton. 2017. *Ideology, Ethics and Policy Development in Public Service Interpreting and Translation.* Bristol: Multilingual Matters.

Wang, Binhua, and Dezheng Feng. 2017. "A Corpus-Based Study of Stance-Taking as Seen from Critical Points in Interpreted Political Discourse." *Perspectives* 26(2): 246–60.

Yanisky-Ravid, Shlomit, and Cynthia Martens. 2019. "From the Myth of Babel to Google Translate: Confronting Malicious Use of Artificial Intelligence – Copyright and Algorithmic Biases in Online Translation Systems." *Seattle University Law Review* 43(1): 99–168.

Zhan, Cheng 詹成. 2012. *Zhengzhi changyu zhong kouyiyuan de tiaokong juese* 政治场域中口译员的调控角色 [The Interpreter's Role as a Mediator in Political Settings]. Beijing 北京: Waiyu jiaoxue yu yanjiiu chuban she 外语教学与研究出版社 [Foreign Language Teaching and Research Press].

8

EXPLORING THE IMPACT OF VIRTUAL AND AUGMENTED REALITY ON INTERPRETING EDUCATION AND PRACTICE

Kan Wu and Dechao Li

8.1 Introduction

The rapid evolution of technology continuously reshapes educational landscapes, introducing novel tools that enhance learning through interactive and immersive experiences. Among these, virtual reality (VR) and augmented reality (AR) stand out due to their ability to create simulated environments in which complex, real-world scenarios can be replicated and interacted with in real time. This chapter explores the integration of VR and AR technologies in the training of interpreting, an activity that requires high levels of cognitive, linguistic and interpersonal skills. The practice of interpreting, bridging language gaps in real time, demands not only fluency in multiple languages but also an acute ability to quickly comprehend, process and relay information accurately and coherently (Pöchhacker 2022). This poses challenges for interpreter training, which must ensure that students are well-prepared to handle the intense and dynamic nature of real-world interpreting situations. Traditional methods often fall short in providing the necessary real-time practice and immediate feedback. Therefore, the integration of VR and AR offers a promising solution by creating immersive scenarios that closely mimic real-life interpreting environments, allowing trainees to hone their skills in a safe and controlled setting.

The training of interpreters has traditionally relied on conventional pedagogical tools that, while effective to an extent, often fall short in preparing students for the unpredictable and dynamic nature of live interpreting. As Angelelli (2006, 23) notes, "Developing cultural sensitivity and cultural responsiveness is a life-long process" that requires a set of skills in interpreting that are difficult to acquire in the artificial environment of the classroom. VR and AR technologies offer promising alternatives to these traditional methods by providing immersive, interactive

DOI: 10.4324/9781003597711-9

experiences that can mimic the pressures and demands of real-life interpreting scenarios without the associated risks and logistics (Akgün and Atıcı 2022). Moreover, VR and AR can significantly enhance the training and practice of interpreters by offering a platform for repeated practice in a controlled yet realistic setting, thereby potentially increasing student engagement and learning outcomes (Kaimara et al. 2022).

The relevance of this chapter is underscored by the ongoing need for innovative educational strategies that can keep pace with the demands of modern professional environments, including the interpreting sector. As global interactions continue to increase, so does the need for competent interpreters. The educational sector must adapt to prepare students more effectively for these roles, and VR and AR technologies represent a significant step forward in this regard.

This chapter explores how VR and AR can be integrated into interpreting education to enhance learning and practice. It will explore the current state of these technologies in educational settings, their potential benefits and the challenges they pose. Furthermore, this chapter will propose methodologies for their implementation and discuss future directions for research and development in this field. By exploring these areas, this chapter seeks to provide a relatively comprehensive overview of the potential of VR and AR to transform the teaching and practice of interpreting, thereby contributing to the broader field of educational technology and its application in specialized domains.

8.2 Background and current practices

Interpreting, vital in facilitating cross-cultural communication, has historically relied on diverse pedagogical strategies, such as lectures, supervised practice sessions and role-playing exercises, aimed at preparing students for the demanding nature of real-time linguistic translation. The evolution of these strategies, from purely theoretical instruction to more interactive, practice-based learning, reflects broader educational trends and the specific needs inherent to interpreting (Gile 2009). Despite these advancements, traditional interpreting training still struggles to fully prepare students for the unpredictable scenarios they will encounter in professional settings. This includes, inter alia, handling high-pressure environments, managing unexpected linguistic or cultural challenges, dealing with technical issues during interpretation and maintaining accuracy and composure under stress.

8.2.1 Historical development of training methods in interpreting

The evolution of training methods in interpreting reflects the shifting demands of the profession and the broader historical contexts in which these changes occurred. Initially, interpreting training was largely informal, predominantly occurring within the contexts of diplomacy and commerce, where apprenticeships were common. In previous interpreter training sessions, mentors, typically experienced interpreters, would guide novices through the nuances of the profession, focusing on direct experience rather than structured classroom learning.

The establishment of formal interpreter training programs did not materialize until the mid-20th century, driven by the burgeoning need for professional interpreters after World War II and the founding of international organizations like the United Nations. This period marked a significant shift toward academic institutionalization of interpreter training, with the first dedicated training programs emerging in Europe, which were soon followed by others around the world (Baigorri-Jalón 2004).

These early programs were heavily influenced by the conference interpreting needs of international diplomacy. They focused on developing a solid foundation in languages and introduced students to the consecutive and simultaneous interpreting techniques that are still fundamental to the profession today. The curriculum was rigorously structured, combining linguistic proficiency with intensive practice sessions designed to simulate real-world interpreting scenarios (Setton and Dawrant 2016).

As the profession evolved, so did the pedagogical approaches. The latter part of the 20th century saw a greater emphasis on research-based education, where empirical studies began informing pedagogical strategies, integrating findings from cognitive science and communication studies into interpreter training (Seeber 2011). This shift not only diversified training methods but also deepened the theoretical underpinnings of interpreter education, balancing practical skills with a robust theoretical framework.

8.2.2 Current pedagogical approaches to interpreting education

Contemporary interpreter training programs are characterized by their structured and diversified approaches, blending theoretical knowledge with practical skills essential for professional practice. These programs typically include a rigorous curriculum that covers linguistic theories, ethics, technological skills and specialized terminology relevant to different interpreting contexts such as legal, medical or conference interpreting (Angelelli 2006). Practical training is a core component, with a strong emphasis on simulating real interpreting situations through role-playing, simulated multilingual conferences and extensive use of peer feedback sessions (Setton and Dawrant 2016).

Advancements in digital technology have further enriched these educational programs. Tools such as virtual classrooms and online glossary databases allow students to practice and refine their skills remotely and asynchronously, accommodating flexible learning schedules and enabling continual skill development outside traditional classroom settings (Fantinuoli 2018). For instance, computer-assisted interpreting tools now play a crucial role in training, offering functionalities like terminology management and real-time access to information, which are indispensable in the fast-paced environment of professional interpreting (Seeber 2017).

Despite the incorporation of these advanced tools, the essence of interpreting education remains the enhancement of real-time decision-making and problem-solving skills, which are critical in live interpreting scenarios. This is achieved not only through technological integration but also through pedagogical strategies that encourage critical thinking, adaptability and ethical decision-making in complex communicative situations.

8.2.3 Current interpreting training tools and their limitations

Current tools in interpreting training have been foundational, but possess inherent limitations that can hinder the development of essential skills needed for professional interpreting. Textbooks, audio recordings and structured classroom lectures are the bedrock of traditional methodologies (Tang and Li 2016, 2017). These tools focus on enhancing linguistic proficiency and providing theoretical knowledge but often lack the interactive component necessary for developing practical interpreting skills (Tang and Li 2017). Audio recordings, for example, are static and unidirectional; they do not allow for the dynamic exchange that characterizes real interpreting situations, nor do they accommodate the interpreter's input in altering the flow of speech or interaction (Riccardi 2005).

Furthermore, the traditional classroom setting is typically unable to replicate the stress and cognitive demands of simultaneous interpreting in a live environment. This creates a gap between the controlled educational environment and the unpredictable nature of real-world interpreting. The inability of traditional tools to provide real-time feedback is another significant shortfall, as immediate correction and guidance are crucial for mastering the rapid decision-making required in professional settings (Seeber 2011).

These limitations highlight the need for more immersive and interactive training tools that can mimic the complex, high-pressure scenarios faced by interpreters. While traditional tools provide a solid theoretical foundation, the evolution of pedagogical strategies toward more engaging and realistic training methods is essential to fully prepare students for the demands of the interpreting profession. The introduction of VR and AR technologies into interpreting education has the potential to greatly enhance current teaching and learning activities. By creating simulated environments that are both immersive and interactive, VR and AR can address the gaps left by traditional methods, providing a more comprehensive training experience that better equips students for professional challenges.

8.3 Benefits of VR and AR in interpreting education

As Pöchhacker (2022, 170) notes, "The use of digital technology to implement various forms of remote interpreting, in conference as well as community-based institutional settings is certain to become an ever more significant". The integration of VR and AR into interpreting education represents a significant advancement in pedagogical strategies within this field. These technologies offer immersive, interactive experiences that mimic the complexities and pressures of real interpreting scenarios, thereby addressing many of the limitations inherent in traditional training methods (Huang et al. 2021). This section explores the various benefits that VR and AR offer to interpreting education, ranging from enhanced engagement to adaptive learning capabilities.

8.3.1 Enhanced engagement and motivation

The immersive nature of VR and AR technologies plays a crucial role in enhancing engagement and motivation among interpreting students. These technologies leverage the sensory-rich environments they create to captivate students' interest and sustain their engagement over time. By simulating real-world interpreting scenarios in a controlled, interactive manner, VR and AR transform the learning experience from passive reception to active participation (Bower 2017). This shift is significant in educational settings where traditional methods may fail to fully engage learners due to their static nature.

Engagement in this context is not merely about capturing attention; it is about fostering a deep connection with the learning material, which facilitates higher levels of cognitive processing and retention. When learners are immersed in a VR or AR environment, they exhibit increased focus and are less likely to be distracted by external factors. Freina and Ott (2015, 5) observed that the immersive nature of VR environments can "increase the learner's involvement and motivation while widening the range of learning styles supported". This heightened focus can lead to a more profound learning experience, as students are not just observing but are part of the scenario, experiencing the tasks and challenges firsthand.

The novelty and technological sophistication of VR and AR are themselves motivating factors. The newness of these tools in educational settings provides a refreshing change from conventional learning modalities, sparking curiosity and interest among students (Dalgarno and Lee 2010). For interpreting students, who must develop a high level of skill in languages and cultural nuances, the ability to practice in diverse, realistic settings can significantly increase motivation. By providing a variety of scenarios – from international conferences to courtroom settings – VR and AR can cater to different learning preferences and needs, further enhancing student motivation.

These immersive technologies support a sense of autonomy and competence, which are critical components of intrinsic motivation according to self-determination theory (Ryan and Deci 2000). In VR and AR settings, students can control their learning pace and choose their learning paths, which empowers them and reinforces their intrinsic motivation to learn and succeed. This autonomy is particularly beneficial in interpreting training, where the ability to handle unpredictable and varied speaking contexts is crucial.

VR and AR technologies significantly improve engagement and motivation by creating immersive, interactive and novel learning environments that encourage active participation and emotional connection to the content. This enhanced engagement not only makes learning more enjoyable but more effective as well, because it promotes deeper understanding and retention of interpreting skills.

8.3.2   Opportunities for simulated practices

VR and AR technologies provide unprecedented opportunities for simulated practice in a range of diverse and complex scenarios that are critical for the training of interpreters. By creating detailed and varied simulated environments, these technologies enable students to experience the nuances and pressures of real interpreting settings without the logistical and financial constraints of organizing live, in-person training sessions (Huang et al. 2021).

The ability of VR to immerse students in realistic 3D environments allows for the simulation of a variety of interpreting contexts, from large international conferences to intimate legal or medical consultations. This exposure is crucial for developing the flexibility and adaptability needed in professional interpreters, who must be able to operate effectively across different cultural and situational contexts (Akgün and Atıcı 2022). For example, VR can replicate the experience of a noisy conference room, providing students with the challenge of maintaining concentration and accuracy in less-than-ideal acoustic conditions, a common real-world challenge. Similarly, AR enhances real-world environments by overlaying digital information, such as texts or live speech transcriptions, directly into the interpreter's field of vision. This feature supports the development of skills such as glossary management under time pressure and improves the interpreter's ability to manage the cognitive load by integrating digital aids seamlessly into the interpreting process (Garzón 2021).

Moreover, these simulated practices are not limited to solo experiences; they can be designed to accommodate multiple participants, thus facilitating group training sessions where students can practice relay interpreting and teamwork. This kind of collaborative interpreting practice is vital, as it mirrors professional situations where interpreters often work in pairs or teams, especially in settings like the European Union or the United Nations.

The flexibility of VR and AR to adapt to the user's performance further enhances their utility in educational settings. These technologies can modify the complexity of scenarios in real time, providing more or less challenging conditions based on the student's proficiency. This adaptive approach ensures that each student can progress at an appropriate pace, gradually increasing their skill level and confidence before they encounter such situations in their professional lives.

8.3.3   Immediate feedback and adaptive learning capabilities

Another advantage of VR and AR is their capacity to provide immediate, personalized feedback to learners. These technologies track user performance and provide real-time corrections and tips, described as "innovative, motivating, and promising" and essential for mastering complex interpreting skills (Hu 2023, 174; Garzón 2021). For example, AR devices can overlay feedback on the user's field of view, pointing out errors in translation or suggesting better phraseology. Imagine a scenario where a student is interpreting a speech in real time using AR glasses. As the student translates, the AR system highlights areas where the interpretation deviates from the source material, offers alternative translations and provides context-specific notes. This overlay of information allows the learner to make immediate adjustments while still engaged in the task, thereby reinforcing proper techniques and improving accuracy. This immediacy helps learners make quick adjustments and internalize corrections effectively, accelerating the learning process. Moreover, VR and AR systems can adapt to the individual learning pace and style of each student, offering a more personalized learning experience. Advanced algorithms analyze performance data to adjust the difficulty level and type of exercises, ensuring that each learner is challenged appropriately and continues to progress (Akgün and Atıcı 2022).

The benefits of integrating VR and AR into interpreting education are manifold and significant. These technologies provide an engaging, immersive and adaptive learning environment that can significantly enhance the preparedness of interpreting students. As the demand for skilled interpreters continues to grow in our increasingly globalized world, the importance of incorporating these advanced technological tools into interpreter training programs cannot be overstated. By doing so, educational institutions can better equip their students with the skills necessary to succeed in this challenging and crucial profession.

8.4 Potential challenges of using VR and AR

While VR and AR technologies present considerable benefits to interpreting education, their integration into academic and professional training environments also comes with significant challenges and considerations. These challenges span technical, pedagogical, ethical and financial domains, each requiring careful consideration and strategic planning to ensure successful implementation.

8.4.1 Technical challenges

The technical challenges associated with integrating VR and AR into interpreting education are significant and multifaceted. One of the primary hurdles is the requirement for sophisticated and often expensive hardware, such as high-performance computers and specialized VR headsets or AR glasses. These devices are necessary to support the complex software applications used in VR and AR training modules, which may represent a substantial financial burden for educational institutions.

Moreover, the software development for VR and AR applications tailored to interpreting training is both resource-intensive and requires a high level of expertise. Developing these applications involves both software engineers and educators and interpreting professionals, so as to ensure that the simulations are accurate and pedagogically effective (Akgün and Atıcı 2022). This multidisciplinary development process can be costly and time-consuming, potentially slowing down the adoption of VR and AR technologies in interpreter training programs.

For example, creating an interpreting simulation that accurately mimics a live conference setting requires detailed programming to handle live speech recognition, translation and real-time feedback, all of which must be seamlessly integrated into a VR environment. This complexity demands a significant investment of time and resources, often beyond the reach of many educational institutions.

Another critical technical challenge is the design of user interfaces (UI) that are intuitive and conducive to learning. A poorly designed UI can lead to user frustration, which can detract from the learning experience and reduce the overall effectiveness of the training. Ensuring that VR and AR applications are accessible and easy to use for all students, including those with disabilities, is essential for their successful integration into educational curricula (Garzón 2021).

Addressing these technical challenges requires substantial upfront investment in both hardware and software, as well as ongoing support to maintain and update the technological infrastructure. This necessitates not only financial resources but also a strategic vision that prioritizes long-term benefits over short-term costs.

8.4.2   Pedagogical challenges

Integrating VR and AR into interpreter training programs introduces several pedagogical challenges that need to be carefully managed. One of the primary issues is ensuring that the use of these technologies aligns with educational objectives rather than being driven by the novelty of the technology itself. There is a risk that the technological aspects could overshadow the pedagogical goals, leading to the use of VR and AR as gimmicks rather than effective educational tools (Dalgarno and Lee 2010).

For instance, an interpreting program might introduce VR simulations of international conferences to provide realistic practice environments. However, if the focus shifts too much toward the impressive visuals and immersive experience rather than on developing critical interpreting skills, the educational value may be compromised. Instructors must ensure that each VR session has clear, targeted learning outcomes that align with the overall curriculum.

Furthermore, the effectiveness of VR and AR in education depends significantly on the instructors' familiarity and comfort with these technologies. Teachers must not only understand how to operate the technology but also how to integrate it effectively into their pedagogical strategies. This requires substantial training and professional development, which can be a barrier to the adoption of VR and AR in educational settings (Lampropoulos et al. 2022).

Another challenge is content creation for VR and AR platforms, which requires a deep understanding of both the technology and the subject matter. Developing immersive and pedagogically sound content that genuinely enhances interpreting skills demands collaboration between technologists, educators and domain experts. This multidisciplinary approach can be resource-intensive and complex, potentially straining the capacities of educational institutions (Akgün and Atıcı 2022).

To overcome these pedagogical challenges, institutions must prioritize clear educational outcomes when designing VR and AR experiences and ensure ongoing support and training for educators. This will help in fully realizing the potential of these technologies to transform interpreter training.

8.4.3 Ethical and accessibility challenges

The deployment of VR and AR in interpreter training raises several ethical and accessibility concerns that must be addressed to ensure equitable and responsible use of these technologies. Ethical issues include the privacy and security of personal data collected during VR and AR sessions. The immersive nature of these technologies means that a significant amount of sensitive user data can be recorded and analyzed, raising concerns about how this data is stored, used and shared (Merchant et al. 2014). Furthermore, there are potential psychological impacts associated with prolonged use of VR, such as disorientation and discomfort, which are ethical concerns that institutions must consider. Ensuring that VR and AR applications do not adversely affect students' mental health is crucial (Kaimara et al. 2022).

Accessibility is another critical area, requiring that VR and AR tools be usable by students with a range of disabilities. This involves designing technologies that accommodate users with visual, auditory or physical impairments, ensuring that no student is disadvantaged by the introduction of new learning modalities (Freina and Ott 2015).

Addressing these ethical and accessibility concerns involves implementing robust data protection policies, designing inclusive and adaptable technologies and maintaining an ongoing dialogue about the ethical implications of immersive learning environments.

8.4.4 Financial challenges

The financial challenges of incorporating VR and AR into interpreter training programs are substantial and multifaceted. Initial costs include the purchase of specialized hardware, such as VR headsets and AR glasses, and the development or acquisition of tailored software applications. These expenses can be prohibitive for many educational institutions, potentially limiting the adoption of these technologies to wealthier, more resource-rich environments (Bower 2017).

Beyond the initial setup costs, there are ongoing expenses related to maintenance, updates, and the training of faculty to effectively use and integrate these technologies into their curricula. Ensuring that VR and AR systems remain up-to-date and continue to meet educational standards requires continuous investment (Huang et al. 2021). The financial burden may also extend to students, particularly if costs are passed down in the form of higher tuition fees or required purchases of personal equipment. This could restrict access to these innovative training tools, potentially widening the gap between different socioeconomic groups within the student population (Akgün and Atıcı 2022).

Addressing these financial challenges requires careful budgeting, potential funding from external grants and a clear analysis of the return on investment, considering both direct educational benefits and long-term professional outcomes for students.

8.5 Methodologies for implementing VR and AR in interpreting training

The integration of VR and AR into interpreter training programs necessitates a methodical approach to ensure effective and sustainable implementation. This section outlines the methodologies for developing VR and AR content, integrating these technologies into interpreting curricula and evaluating their effectiveness and discusses scalability and customization options.

8.5.1 Development of VR and AR content

Developing content for VR and AR in interpreter training involves a multidisciplinary approach that brings together expertise from technology developers, educators and interpreting professionals. The first step in this process is the identification of specific training needs that VR and AR can address more effectively than traditional methods. This might include complex interpreting scenarios that require high levels of immersion or tasks that benefit from interactive elements and immediate feedback (Huang et al. 2021).

Following the needs analysis, the design phase involves creating detailed and accurate simulations of interpreting environments. For VR, this could mean developing three-dimensional recreations of conference halls or courtrooms where students can practice simultaneous or consecutive interpreting. In AR, this might involve designing overlays that provide students with real-time contextual information or translation aids during practice sessions, enhancing their learning experience and response accuracy (Garzón 2021).

The success of these initiatives largely depends on the input from experienced interpreters to ensure that the scenarios are realistic and relevant. Additionally, working closely with educational technologists is crucial to integrate pedagogical objectives effectively and ensure the technology enhances learning rather than distracting from it. This collaborative approach helps in creating VR and AR content that is not only technologically advanced but also pedagogically sound and directly tailored to the needs of interpreting students.

8.5.2 Integration into interpreting courses

Integrating VR and AR technologies into interpreting courses requires careful planning and curriculum design. Educators must align VR and AR modules with the learning objectives of the course and ensure they complement rather than replace traditional teaching methods. This integration should be designed to enhance practical skills, such as quick decision-making, accuracy under pressure and adaptability to different interpreting modes and settings (Akgün and Atıcı 2022).

Faculty training is also a critical component of integration. Instructors must be trained not merely in the technical use of these tools, but also in how to effectively incorporate them into their existing curricula. This might involve instructional design support to help faculty develop new teaching modules that make the best use of VR and AR for learning specific interpreting skills (Lampropoulos et al. 2022). It is important to ensure that the technology is seamlessly embedded into the course structure. This may require adjustments in course timelines and assessments to accommodate new types of learning activities. For instance, VR sessions may be used to simulate interpreting assignments and subsequent classes can be used to debrief and discuss the experiences and learning outcomes (Class and Moser-Mercer 2013).

The integration process also needs to consider student access to technology, ensuring all students have equal opportunities to engage with VR and AR resources. This might include setting up dedicated VR labs or providing portable AR devices, depending on resource availability.

8.5.3 Evaluation of effectiveness

To assess the effectiveness of VR and AR in interpreter training, comprehensive evaluation methodologies need to be established. This should include both qualitative and quantitative measures to provide a comprehensive overview of their impact on student learning and skill development.

Quantitative data can be gathered through performance metrics derived from the VR and AR systems themselves, which often have built-in analytics to track user progress, error rates and completion times. This data provides objective evidence of improvements in specific interpreting skills, such as reaction time and accuracy under simulated conditions (Huang et al. 2021). Qualitative feedback is equally important and can be obtained through surveys, focus groups and interviews with students and faculty. These methods allow participants to express their perceptions of the VR and AR learning experience, including its usability, engagement and overall impact on their educational journey (Lampropoulos et al. 2022). Additionally, it is beneficial to include self-assessment techniques where students reflect on their own learning and perceived improvements, providing insights into their confidence and professional preparedness. Longitudinal studies can also play a vital role in this evaluation, tracking students' performance over time to assess the long-term benefits of VR and AR training. Such studies help determine whether skills acquired in virtual environments translate into real-world competencies, a critical factor for professional interpreting (Garzón 2021).

The combination of these diverse evaluation methods helps in accurately assessing the effectiveness of VR and AR technologies in interpreter training, guiding future improvements and adoption strategies.

8.5.4 Scalability and customization

For VR and AR technologies to be truly impactful in the field of interpreter training, they must be scalable and customizable. Scalability ensures that VR and AR technologies can be expanded and adapted to meet the needs of a growing number of users across different geographic and institutional contexts without a loss in functionality or educational value. Cloud-based solutions are particularly effective in this regard, as they allow institutions to scale their use of VR and AR resources according to demand, providing cost-effective access to high-quality educational tools without the need for significant hardware investments (Bower 2017).

Customization is equally important, as it allows VR and AR applications to be tailored to meet the specific learning objectives and challenges of interpreter training. This includes adapting scenarios to different interpreting contexts – such as legal, medical or conference interpreting – and accommodating various language pairs and cultural nuances. Customizable content ensures that students can engage in meaningful practice that directly aligns with their educational and professional goals (Akgün and Atıcı 2022). Moreover, customization can address individual learning styles and paces, enhancing the learning experience by allowing students to focus on areas where they need the most practice or advancing more quickly through areas where they excel. This level of personalization is crucial for maximizing learning outcomes and ensuring that all students can benefit from the use of these technologies (Class and Moser-Mercer 2013).

To effectively implement scalable and customizable VR and AR solutions, institutions must invest in flexible platforms that support easy updates and modifications. This would not only extend the lifespan of the educational tools but also keep content relevant and aligned with the latest pedagogical and technological advancements.

8.5.5 Examples of implementing VR and AR

The implementation of VR and AR in interpreting training has been explored through various case studies and pilot programs across educational institutions worldwide. These examples provide valuable insights into the practical applications of these technologies and their impact on enhancing interpreter training.

A notable example is the integration of the VR technology into a conference interpreting course. In this case, VR simulations were used to mimic international conferences, providing students with a platform to practice simultaneous interpreting in a highly realistic and immersive environment. The VR setup included virtual audiences, multiple live speakers and real-time feedback mechanisms. Significant improvements are expected to be found in students' ability to handle live interpreting scenarios, particularly in managing stress and cognitive load. Students would also have increased confidence and better preparation for real-world interpreting tasks (Huang et al. 2021).

Another example is the use of AR in medical interpreting courses. Here, AR technology was employed to overlay a simulated doctor-patient interaction over the real environment, allowing students to practice medical interpreting with virtual characters who mimicked patients and healthcare professionals. This AR application would provide visual and textual cues to help students manage medical terminology more effectively, enhancing their accuracy and speed in translation. It is expected that AR would help improve students' performance in medical interpreting, particularly in terms of terminology accuracy and procedural understanding (Garzón 2021).

Comparing these VR and AR applications with traditional teaching methods reveals several advantages. Traditional methods often rely on role-plays and audio-visual materials that do not fully replicate the dynamic nature of interpreting scenarios. In contrast, VR and AR provide a more authentic and engaging experience, offering realistic practice opportunities and immediate feedback, which are crucial for developing practical interpreting skills. Moreover, VR and AR allow for repeated practice in a controlled environment, where students can learn from their mistakes without the real-world consequences. This aspect of VR and AR is particularly beneficial in interpreting training, where the ability to handle diverse and challenging scenarios is paramount.

These existing examples demonstrate that VR and AR have the potential to significantly enhance the quality and effectiveness of interpreter training programs. By providing immersive, realistic and interactive learning experiences, these technologies prepare students more effectively for the demands of professional interpreting.

8.6 Future directions and research

As VR and AR technologies continue to evolve, their integration into interpreter training presents new opportunities and challenges. This section discusses potential advancements in VR and AR technology, emerging trends in educational technology, and identifies key areas for future research and experimental designs that could further enhance interpreting education.

8.6.1 Advancements in VR and AR technology

The continuous advancement of VR and AR technologies holds promising potential for further enhancing interpreter training programs. One of the key areas of development is the incorporation of multisensory feedback systems into VR environments. These systems could provide tactile feedback (haptic technology), olfactory cues and more advanced auditory feedback, creating a fully immersive and sensory-rich learning experience. Such enhancements would be particularly beneficial in scenarios where physical sensations and environmental cues play a crucial role in communication, such as in medical or conflict-zone interpreting (Bower 2017).

The integration of advanced artificial intelligence (AI) within VR and AR platforms is poised to revolutionize how feedback is delivered and how student performance is assessed. AI algorithms could analyze complex patterns of student behavior, offering more personalized and adaptive learning experiences. This technology could potentially identify subtle nuances in student responses, adjust scenarios in real time and provide targeted feedback based on individual performance metrics. Moreover, the future of VR and AR may see greater cloud integration, which would facilitate easier updates, scalability and access across different devices and locations. Cloud-based VR and AR could democratize access to high-quality interpreter training resources, making them more widely available to institutions regardless of their geographical location or economic status (Garzón 2021).

These technological advancements could significantly refine the quality and effectiveness of interpreter training, creating more nuanced, realistic and adaptable educational environments.

8.6.2   *Emerging trends in educational technology*

The integration of VR and AR with other emerging technologies could revolutionize how educational content is delivered and authenticated. One such trend is the integration of machine learning (ML) algorithms with VR and AR systems. This integration facilitates adaptive learning environments that can dynamically adjust to the skill level and learning pace of individual students. ML can analyze vast amounts of data from student interactions within VR and AR settings to optimize the training modules in real time, enhancing personalized learning experiences and improving linguistic and interpretive skills more efficiently (Garzón 2021).

Blockchain technology is another emerging trend with the potential to transform educational settings, including interpreter training. By utilizing blockchain, educational institutions can create secure, immutable records of student achievements and certifications. This technology ensures the authenticity and portability of qualifications, making it easier for interpreting professionals to prove their credentials and for employers to verify them, thereby enhancing professional mobility and transparency in the field (Bower 2017).

The shift toward cloud-based VR and AR applications is significant. Cloud computing offers the potential to lower the barriers to entry for adopting these technologies by reducing the need for expensive hardware and facilitating easier updates and maintenance. This could enable a broader range of institutions to implement advanced VR and AR training solutions, thereby democratizing access to cutting-edge educational tools (Akgün and Atıcı 2022).

These trends indicate a move toward more accessible, personalized and secure educational technologies in interpreter training, promising to enhance both the quality and reach of interpreter education.

8.6.3 Recommendations for future research

To maximize the benefits of VR and AR in interpreter training, several areas require further research:

Effectiveness of VR and AR in varied interpreting contexts: Studies should explore the effectiveness of VR and AR across different interpreting specialties, such as legal, medical or community interpreting. Research could examine how different features of VR and AR can be customized to address the unique challenges of these fields (Class and Moser-Mercer 2013).

Longitudinal impact of VR and AR training: Long-term studies are needed to assess the sustained impact of VR and AR training on professional interpreting practice. This research could track career progression, job satisfaction and professional competence over several years (Freina and Ott 2015).

Accessibility and inclusivity in VR and AR applications: Further research is required to ensure that VR and AR tools are accessible to all students, including those with disabilities. Studies could focus on developing inclusive design practices that accommodate a wide range of physical and cognitive abilities (Garzón 2021).

Integration of VR and AR into interpreting certification programs: Investigating how VR and AR training can be formally recognized in certification and accreditation processes for interpreters. This would help establish standards and benchmarks for VR and AR in professional training curricula.

The future of interpreting education through VR and AR technologies is promising but requires concerted efforts in research and development to realize its full potential. By addressing the outlined future directions and research needs, stakeholders in interpreting education can ensure that these technologies would enhance learning experiences and contribute to the professional readiness and ongoing development of interpreters.

8.7 Conclusion

The integration of VR and AR into interpreter training marks a significant evolution in the pedagogical approach to educating future professionals in this field. As outlined in this paper, these technologies offer immersive, interactive and highly adaptable training environments that can significantly enhance the learning experience and outcomes for interpreting students. From increasing engagement and motivation to providing opportunities for simulated practice in diverse and complex scenarios, VR and AR technologies transform traditional interpreter training methods by addressing their limitations and expanding their capabilities.

VR and AR allow for the simulation of real-world interpreting environments in a controlled, risk-free setting, offering students the unique opportunity to practice and hone their skills in a variety of complex situations. The immediate feedback and adaptive learning capabilities of these technologies further empower students to improve their performance dynamically and efficiently. These benefits are not just theoretical but are supported by case studies and examples from institutions that have successfully integrated VR and AR into their curricula, showing marked improvements in student competency and readiness.

However, the implementation of these technologies is not without its challenges. Technical, pedagogical, ethical and financial considerations must be carefully managed to ensure that the adoption of VR and AR tools achieves the desired educational outcomes. Addressing these challenges requires a concerted effort from educational institutions, technology developers and policymakers to create environments that are technically robust, pedagogically sound, ethically governed and financially viable.

Looking to the future, the field of interpreter training is poised to continue its technological evolution. Advancements in VR and AR, along with the integration of other emerging technologies such as AI and ML, promise to further enhance the effectiveness and accessibility of interpreter training. Research into the long-term impacts of these technologies on professional practice, the development of new content and the exploration of innovative uses will be critical. As this technology evolves, so too must the methodologies for its application in educational settings, ensuring that it serves to enhance pedagogical practices and meet the evolving needs of the interpreting profession.

In conclusion, VR and AR represent transformative tools for interpreter training, providing powerful new ways to meet the demands of this challenging and crucial field. The ongoing innovation and research into these technologies will continue to shape the landscape of interpreter education, making it more dynamic, effective and aligned with the complexities of global communication. As we move forward, it is essential that educators, researchers and technologists work together to harness the potential of VR and AR, ensuring they are used to their fullest potential in training the next generation of interpreters.

References

Akgün, Muhterem, and Bünyamin Atıcı. 2022. "The Effects of Immersive Virtual Reality Environments on Students' Academic Achievement: A Meta-Analytical and Meta-Thematic Study." *Participatory Educational Research* 9(3): 111–31.

Angelelli, Claudia V. 2006. "Validating Professional Standards and Codes: Challenges and Opportunities." *Interpreting* 8(2): 175–93.

Baigorri-Jalón, Jesús. 2004. From *Paris to Nuremberg: The Birth of Conference Interpreting*. Amsterdam/Philadelphia: John Benjamins Publishing Company.

Bower, Matt. 2017. *Design of Technology-Enhanced Learning: Integrating Research and Practice*. Bingley: Emerald Publishing Limited.

Class, Barbara, and Barbara Moser-Mercer. 2013. "Training Conference Interpreter Trainers with Technology–A Virtual Reality." *Quality in Interpreting: Widening the Scope* 1: 293–313.

Dalgarno, Barney, and Mark J. W. Lee. 2010. "What Are the Learning Affordances of 3-D Virtual Environments?" *British Journal of Educational Technology* 41(1): 10–32.

Fantinuoli, Claudio. 2018. *Interpreting and Technology*. Berlin: Language Science Press.

Freina, Laura, and Michela Ott. 2015. "A Literature Review on Immersive Virtual Reality in Education: State of The Art and Perspectives." The International Scientific Conference eLearning & Software for Education Conference.

Garzón, Juan. 2021. "An Overview of Twenty-Five Years of Augmented Reality in Education." *Multimodal Technologies and Interaction* 5(7): 37.

Gile, Daniel. 2009. Basic *Concepts and Models for Interpreter and Translator Training*. Amsterdam/Philadelphia: John Benjamins Publishing Company.

Hu, Peiyang. 2023. "Business Interpreter Training in the 3D Virtual Reality Environment: A Pilot Study in China." *Scholars International Journal of Linguistics and Literature* 6(3): 169–76.

Huang, Xianlin, Di Zou, Guanxiang Cheng, and Haoran Xie. 2021. "A Systematic Review of AR and VR Enhanced Language Learning." *Sustainability* 13(9): 4639.

Kaimara, Polyxeni, Andreas Oikonomou, and Ioannis Deliyannis. 2022. "Could Virtual Reality Applications Pose Real Risks to Children and Adolescents? A Systematic Review of Ethical Issues and Concerns." *Virtual Reality* 26(2): 697–735.

Lampropoulos, Georgios, Efthymios Keramopoulos, Konstantinos Diamantaras, and Georgios Evangelidis. 2022. "Augmented Reality and Virtual Reality in Education: Public Perspectives, Sentiments, Attitudes, and Discourses." *Education Sciences* 12(11): 798.

Merchant, Zahira, Ernest T. Goetz, Lauren Cifuentes, Wendy Keeney-Kennicutt, and Trina J. Davis. 2014. "Effectiveness of Virtual Reality-Based Instruction on Students' Learning Outcomes in K-12 and Higher Education: A Meta-Analysis." *Computers & Rducation* 70: 29–40.

Pöchhacker, Franz. 2022. Introducing *Interpreting Studies*. London/New York: Routledge.

Riccardi, Alessandra. 2005. "On the Evolution of Interpreting Strategies in Simultaneous Interpreting." *Interpreting* 7(1): 55–72.

Ryan, Richard M., and Edward L. Deci. 2000. "Self-Determination Theory and the Facilitation of Intrinsic Motivation, Social Development, and Well-Being." *American Psychologist* 55(1): 68–78.

Seeber, Kilian G. 2011. "Cognitive Load in Simultaneous Interpreting: Existing Theories – New Models." *Interpreting* 13(2): 176–204.

Seeber, Kilian G. 2017. "Multimodal Processing in Simultaneous Interpreting." In *The Handbook of Translation and Cognition*, edited by Ofelia García, Akira Apter, and Miriam Shlesinger, 461–75. Hoboken: Wiley-Blackwell.

Setton, Robin, and Andrew Dawrant. 2016. Conference *Interpreting: A Complete Course and Trainer's Guide*. Amsterdam/Philadelphia: John Benjamins Publishing Company.

Tang, Fang, and Dechao Li. 2016. "Explicitation Patterns in English-Chinese Consecutive Interpreting: Differences Between Professional and Trainee Interpreters." *Perspectives* 24(2): 235–55.

Tang, Fang, and Dechao Li. 2017. "A Corpus-Based Investigation of Explicitation Patterns Between Professional and Student Interpreters in Chinese-English Consecutive Interpreting." *The Interpreter and Translator Trainer* 11(4): 373–95.

9

THE ROLE OF ELECTRONIC CORPORA IN INTERPRETER TRAINING AND EDUCATION

A comprehensive review

Cui Xu

9.1 Introduction

Since its introduction to translation studies by Baker (1993, 1995), the utilization of electronic corpora has become an essential skill in translator training and education. This trend has been further emphasized with the advancement of technology, with scholars highlighting "instrumental competence" (Kelly 2005; PACTE Group 2003; Rodríguez-Inés and Albir 2012), "information mining competence" (EMT Expert Group 2009) or "technological competence" (Laviosa and Falco 2023) as crucial components of translation competence. Over the past two decades, significant academic efforts have been directed toward integrating corpus tools and various types of corpora into translation classrooms, exemplified by the works of Zanettin, Fantinuoli and Bowker.

However, while the impact of electronic corpora on translator training is well-documented, its application in interpreter training and education remains relatively understudied and underutilized. This discrepancy is not surprising, given the inherent challenges associated with data accessibility and corpus compilation in interpreting studies (Shlesinger 2009). Nevertheless, with the impending "technological turn" in the translation and interpreting profession (Fantinuoli 2018b), it is imperative for interpreters to possess instrumental competence, including the ability to mine corpus data and utilize corpus tools, to effectively navigate this evolving landscape. To attain this goal, educators must first underscore the crucial role of electronic corpora in interpreting before incorporating corpus-based activities into their teaching.

This chapter seeks to address the above-mentioned gap by exploring the role and practical applications of corpora and corpus tools in interpreter training and education. While acknowledging the extensive research conducted in the field of translator

DOI: 10.4324/9781003597711-10

training, we aim to leverage insights from these studies to inform and enhance interpreter training practices. In the following sections, we begin by presenting an overview of the different types of corpora as well as the corpus tools commonly used in translator training and education. Subsequently, we move on to the current theoretical discussions and practical applications of corpus tools and resources in interpreter training and education, before presenting some innovative approaches and emerging trends in this field. We then proceed to discuss the possible challenges and ways forward to further inform interpreting practices and the whole profession.

In essence, this chapter serves as a comprehensive revisit of the role and significance of corpora and corpus tools in interpreter training and education. By bridging the gap between translator and interpreter training, we aim to contribute to the ongoing dialogue surrounding the integration of corpus linguistics in interpreter education, highlighting the necessity and importance of incorporating corpora in interpreter training programs.

9.2 Corpora in translator training and education

9.2.1 Types of corpora and related studies

Electronic corpora have become indispensable tools in translator training and education, offering valuable resources for translation practice, research and pedagogy. By providing access to authentic language data, electronic corpora enable translators to develop their language proficiency, translation skills and domain knowledge (Bowker and Pearson 2002).

Different types of electronic corpora, each serving specific purposes, have been utilized in translation pedagogy. One of the most widely accessible and utilized type is monolingual corpora. Serving as a valuable resource for understanding the nuances and usage patterns of a single language (often the target language), this type of corpora can help translators reduce the unwanted "shining through" (Teich 2003) of the source text in the target text, thus helping them produce more natural-sounding translations (Bernardini et al. 2003). However, monolingual corpora are not without their limitations, especially in terms of translation teaching. While they provide authentic examples and contextualized usage in the target language, they do not present ready translation equivalents. Therefore, translators need to consult additional sources to make better translation choices, of which parallel corpora serve as a useful one.

Parallel corpora consist of texts that are translated into one or more languages, usually aligned at the sentence or phrase level. They are crucial for comparative analysis and for understanding how different languages handle specific translation challenges (Bernardini et al. 2003). Translators can use parallel corpora to study translation equivalents or strategies, observe how particular terms or expressions are translated across languages and identify patterns of translation that can inform their work. For example, Liu et al. (2023) explored the pedagogical potentials of parallel corpus in translation training between Chinese and English. Their experimental study showed that students who leveraged parallel corpus outperformed their counterparts who relied on other tools and resources other than corpora in the English-Chinese translation direction, but not the other way around, suggesting that "the parallel corpus had a more beneficial effect on translation into the students' native language compared to translation out of the native language" (144). Overall, their study illustrated the effectiveness of parallel corpora in translation practice and teaching. Despite this, it is also suggested that parallel corpora be used in combination with other types of corpora or resources, because each type provides different aspects of information that can complement with each other and better improve translation performance. However, compared to monolingual corpora, parallel corpora are more difficult to access. The most widely acknowledged parallel corpora include the European Parliament Translation and Interpreting Corpus (EPTIC) project (Bernardini et al. 2016), which provides a large collection of both translated and interpreted texts aligned at the sentence level.

The third type of corpora is comparable corpora. Comprising texts in two or more languages that are not direct translations of each other but are similar in genre, style and subject matter, this type of corpora plays an important role in translation practice and teaching (Laviosa and Falco 2023; Liu 2020; Liu et al. 2023; Zanettin 1998). In fact, Laviosa and Falco (2023) highly acclaimed the role of comparable corpora by stating that the distinctiveness about "research into corpus-based translator education […] is the use of bilingual comparable corpora or monolingual target language corpora as sources of data for experimental or classroom-based observational studies" (p. 10). Biel (2017) proposed two methods for using them in the translation classroom, including (1) use of corpora to translate so that translators can get some hands-on experience during the translation process and (2) study of corpora to reflect upon the translation process. Both methods were proven to be effective in her training programme. Additionally, comparable corpora can also "help learners investigate the respective expectations, experience and knowledge of the linguistic communities involved", thus facilitating them in engaging in "meaning-creation activity and develop[ing] procedural skills" (Zanettin 1998, 3–4).

A fourth type of corpora focusing on specific fields or domains is called specialized corpora. Different from general types of corpora, they provide in-depth insights into the terminology, phraseology and discourse conventions of specialized areas. These corpora are thus essential for translators working in specialized fields, such as economics, technology and law, offering reliable and context-specific examples that help ensure accuracy and appropriateness in terminology and style (Bernardini et al. 2003). Previous studies exploring the use of specialized corpora, often combined with other corpus types such as comparable corpora and learner corpora (as will shortly be discussed), in translation classrooms include Biel (2011, 2017) on legal translation, Bowker (2000) and Bowker and Pearson (2002), among others.

With the growing awareness of interdisciplinary collaboration, learner corpora have been receiving increasing academic attention over the past two decades. Being collections of texts produced by language learners, they are used to study the linguistic patterns, errors and development of learners at various proficiency levels. In the context of translation studies, learner corpora can provide insights into how translation students develop their skills and where they commonly encounter difficulties (Granger 2013). Additionally, they can assist educators track the progress of translation students over time, helping to understand how their translation skills evolve (Bowker and Bennison 2003). Besides, learner corpora can also be used in the classroom to provide authentic examples of learner translations, which can be analyzed and discussed to improve learning outcomes (Monzó 2003).

One more type of corpora needs to be introduced is DIY (Do-It-Yourself) corpora (also called disposable corpora), compiled by individuals such as translators or translation students for specific purposes. These corpora are often built from scratch using texts that are relevant to a particular project or area of study, and their construction and compilation require a thorough understanding of corpus linguistics as a methodology in the translation classroom, which puts higher demand on translation trainers and educators. Nevertheless, this type of corpora provides highly customized resources for translation needs, which is very conducive to the learning and practice of translation.

The corpora introduced above have been experimented with or even incorporated into different translation training programs, as reported in Biel (2017), Liu et al. (2023), Obrusnik (2023), Simonnæs et al. (2023) and Zanettin (2009), to name just a few. In addition to providing access to authentic data, perhaps the greatest pedagogical value of these corpora, as rightfully contended by Bernardini et al. (2003), lies in their "thought-provoking" rather than "question-answering" potentials. Both translation trainers and trainees can explore these corpora to their full potentials and reflect upon their own translation teaching and practices.

9.2.2 Corpus tools and applications commonly used

Corpus data cannot be processed and analyzed without the assistance of corpus tools, i.e., software applications which allow users to explore linguistic patterns, frequencies, collocations and other textual features crucial for various linguistic (and translation-related) tasks (McEnery and Hardie 2012). These tools support tasks, such as concordance searches, keyword analysis and text alignment, which are essential for understanding and translating texts accurately.

Two commonly used standalone corpus tools are AntConc (Anthony 2004, 2005) and WordSmith (Scott 2008). They are particularly useful for DIY corpora, as users can upload their textual data and carry out relevant analysis tasks. For example, Rodríguez-Inés and Albir (2012), in their study assessing students' instrumental competence in using electronic corpora, first introduced the functions and usages of WordSmith to the students before proceeding to the action research and reported positive feedback from the students with respect to this toolkit in developing students' learning autonomy. Similarly, AntConc has been used in translator training to inform either students or lecturers in the translation classroom (Laursen and Pellón

2012). However, it should be noted that both AntConc and WordSmith support only single-language search tasks. If translators want to explore bilingual texts, they need to resort to other corpus tools which support text alignment, such as ParaConc.

Developed by Michael Barlow (1995), ParaConc is a bilingual or multilingual Windows concordancer that can be used in contrastive language learning and translation studies and training. It features functions such as text alignment, collocations and frequencies of collocates. Moropa (2007), for example, explored the English-Xhosa parallel corpus of technical texts using ParaConc. His study showed that the tool was effective in identifying equivalent technical terms and their usage in both languages, thereby enhancing the accuracy of translations. Bowker and Barlow (2008) compared ParaConc, a less well-known corpus tool in the translation industry, with the more widely applied translation memory systems such as SDL Trados, and found that both tools have their strengths and limitations and they can be used in a complementary manner to better facilitate translation practice. Their study, however, also suggests a gap between translation research and practice, as corpus tools are more widely acknowledged and explored in translation research than in translation pedagogy.

A more recent tool widely used in linguistic analysis is #LancsBox (Brezina et al. 2018), which offers advanced visualization options and user-friendly interfaces, in addition to a range of functions as provided by other corpus tools, such as collocation analysis, frequency lists and keyword extraction. However, empirical studies exploring the usefulness of #LancsBox in translator training and education are far and few, suggesting a gap between "corpus-assisted translation teaching" (Liu 2020; Moratto 2023) and that of language teaching. On a further note, it also reflects the under-exploration of corpus-based methodology in translation pedagogy.

In addition to standalone software applications, there are some web-based corpus tools that have proven to be effective in translation teaching. One representative example is Sketch Engine (Kilgarriff et al. 2008), widely used in both professional and educational contexts. It offers functionalities such as corpus building and management, a large variety of readily available corpora, thesaurus and word sketches, in addition to the traditional collocations and keyword analysis functions, making it a versatile resource for translators. Buendía-Castro and López-Rodríguez's (2013) experimental study demonstrated the effectiveness of Sketch Engine as a corpus tool in the construction of disposable corpora directly from the web by the students, compared to traditional methods of manual corpus compilation, thus highlighting the unique features of web-based corpus tools in translation teaching.

9.3 Corpora in interpreter training and education

9.3.1 Theoretical discussions

As introduced above, discussions and investigations of the role of electronic corpora in translator training are rigorous. However, those for interpreter training have remained underexplored. By searching such keywords as "corpus/corpora in interpreter training", "corpus/corpora in interpreter training and education" and "corpus/corpora in interpreting classroom" in Translation Studies Bibliography and filtering out research-oriented studies, we found only a very limited number of theoretical and empirical discussions on the role of electronic corpora in interpreting pedagogy.

Theoretical discussions on the usefulness of corpora in interpreter training started only about a decade ago, often following significant discussions on the exploitation of corpora in interpreting research. Russo (2010), reflecting upon interpreting practice at the time, suggested that students could study professional interpreters' performances to detect strategies, language-pair patterns or other relevant features using the European Parliament Interpreting Corpus (EPIC) Multimedia Archive and the EPIC corpus. These resources have proven to be excellent materials for students' interpreting practice. Bendazzoli (2010) expanded on this by highlighting the "advantages and limitations of resorting to the European Parliament (EP) as a source of material to study (and teach) SI" (p. 51). He noted that students can leverage the real-life performances of professional interpreters to improve their interpreting quality. However, he also cautioned that heavy reliance on the EP resources might limit student interpreters' creativity. Thus, both students and trainers need to strike a balance in how to utilize corpora most efficiently.

Following the same line of thought, Sandrelli (2010) made a modest proposal by arguing for the usefulness of electronic corpora in interpreter training. Similarly, Gorjanc (2011) proposed the idea of corpus-driven interpreter training, calling for the exploitation of the internet as a wealth of specialized language data (or "corpora") and the inclusion of these resources in the educational process.

A more recent discussion on the role of electronic corpora in interpreter training has been undertaken by Fantinuoli (2018b), who discussed two types of corpora particularly useful for interpreting: comparable corpora and ad hoc specialized corpora. He emphasized the potential of these corpora to enhance the preparation phase before interpreting assignments. Fantinuoli (2018b) argued that comparable corpora can help interpreters prepare by familiarizing themselves with subject-specific terminology and phraseology. Specialized corpora, on the other hand, can provide in-depth insights into the specific language used in particular fields, such as medical or legal interpreting.

This short review reveals that, compared to theoretical discussions on the role of electronic corpora in translator training, those in interpreting has been far lagging behind, especially concerning the exploitation of various types of corpora. For example, the potential of learner corpora and disposable corpora have been rarely discussed, along with the underutilization of corpus tools. Fortunately, relevant empirical studies have begun to address this gap by highlighting the potential benefits offered by, for example, learner corpora, as will be examined in Section 9.3.2.

9.3.2 Empirical studies

Apart from the limited theoretical discussions, some scholars (Bale 2013, 2015; Pan et al. 2022; Spinzi 2017; Xu 2018) begun to explore the role of electronic corpora in a number of observational or experimental studies to demonstrate the necessity of corpus-assisted interpreting pedagogy.

The first empirical study exploring the role of electronic corpora in interpreter trainer can be traced back to Manuel Jerez (2006). Jerez investigated the means offered by new technologies, specifically video corpora consisting of real-life inter-preting scenarios, to bring interpreter training closer to real-life communicative sit-uations. Combining observation studies with action research, Jerez found that the introduction of video corpora as training materials increased student motivation dur-ing their learning process. He thus called for interpreting trainers to make full use of these new resources from the very early stages of specialized interpreter training.

Later, Bale (2013, 2015) explored the usefulness of certain spoken multime-dia corpora named BACKBONE, pedagogic corpora for content and language-integrated learning, in undergraduate consecutive interpreting (CI). In his earlier pilot study, Bale (2013) demonstrated the effectiveness of corpus-based exercises in improving students' interpreting competence as well as their foreign language competence in terms of lexical knowledge. His follow-up study (2015) further expanded the experiment by including four case studies and a follow-up ques-tionnaire survey evaluating the efficiency of corpus-assisted interpreting teach-ing. The results showed generally positive attitudes toward the incorporation of electronic corpora, particularly exploited in language-based and interpreting-based exercises. One interesting finding of Bale (2015) was the revelation of students' methods of using the corpora during self-study, which can offer valuable insights into self-learning.

In addition to the exploration of spoken corpora in traditional SI and CI settings, other scholars have charted a different course by introducing electronic corpora in dialogue interpreting settings. Spinzi (2017), for example, explored the role of parallel and monolingual corpora in public service interpreting in the legal sector. To achieve this purpose, she constructed a computerized learner corpus consisting of legal texts and compared it to both monolingual and parallel corpora to identify the stumbling blocks dialogue interpreters struggle with, such as discourse mark-ers and phraseological constructions. Her study demonstrated the necessity and usefulness of corpus tools and resources in facilitating both academic research and Public Service Interpreting and Translation (PSIT) practitioners. Focusing on a similar scenario, Dal Fovo (2018) addressed the teaching of dialogue interpreting in healthcare settings by introducing real-life data from the Healthcare Interpreting Quality Corpus (HIQC) project to her English-Italian healthcare interpreting class-room, in addition to the traditional role-playing method. Her study suggested that the introduction of real-life interpreting data significantly improved students' understanding and handling of real-world scenarios, emphasizing the practical applications of corpus data in interpreter training.

A slightly different exploitation of interpreting corpora was done by Xu (2018), who tested the usefulness of corpus-based terminological preparation in enhanc-ing student interpreters' accuracy in Chinese-English simultaneous interpreting and vice versa. Xu introduced two corpus tools, including a term extraction tool and a concordance tool, to facilitate the preparation process. The results showed that students who used these tools demonstrated better terminological accuracy and performance of interpreting quality, and the subsequent recall of terms was also better for this group, thus underscoring the value of corpus-based resources in interpreter training.

In a more recent study, Pan et al. (2022) highlighted the role of learner corpora in translator and interpreter training. Drawing on self-constructed learner corpora – the Chinese/English Translation and Interpreting Learner corpus (CETILC) developed for the study of lexical cohesion, they introduced three case studies to explore the potentials of learner data through different annotation methods: human annotation, machine-facilitated human annotation and human-supervised/edited machine annotation. Their findings highlighted the complexity of learner language and the benefits of using learner corpora to tailor training programs to address specific areas of difficulty for students.

9.3.3   Exploration of innovative approaches and emerging trends

As the field of interpreter training continues to evolve with technological advancements, innovative approaches and emerging trends are gradually shaping the landscape of corpus-assisted interpreting teaching. Recent developments in technology and pedagogy offer new opportunities for enhancing interpreter training, building upon the solid foundation established by traditional empirical studies and theoretical discussions in both translator and interpreter training.

One significant trend is the integration of computer-assisted interpreting (CAI) tools and technologies. Fantinuoli (2018a) discussed three major interpreting-related technologies: CAI, remote interpreting and machine interpreting (MI). Although his discussion was set against the broader context of the interpreting profession instead of corpus-assisted interpreter training, we believe these tools can help enhance the functionality of traditional corpora and provide interpreters with sophisticated resources to improve their work experience. Specifically, CAI tools extend the functionality of traditional corpora by enabling dynamic glossary generation and topic identification, thereby equipping interpreters with decision-making support. Concurrently, machine learning algorithms applied to large corpora can reveal latent linguistic patterns and trends – insights that not only advance MI systems but also inform the design of evidence-based pedagogical interventions.

Another emerging trend in interpreting pedagogy, as reviewed in Section 9.3.2, is the introduction of blended learning approaches that combine traditional face-to-face instruction with digital resources (such as corpora) and online learning. Dal Fovo (2018), for example, demonstrated the benefits of blended learning approaches by integrating corpus resources (i.e., simulated interpreting scenarios) into traditional interpreting classroom, making interpreter training more adaptive and responsive to individual learner needs. However, to guarantee the effectiveness of such blended learning approaches, a well-designed curriculum is essential, requiring closer collaboration between interpreting trainers and researchers, as will be discussed in Section 9.4.2.

9.4   Bridging translator and interpreter training through corpus-based methodologies

9.4.1   Challenges for corpus-assisted interpreter training

As reviewed above, exploration of the potentials of electronic corpora in interpreter training has been far lagging behind compared to that in the translation pedagogy, especially with regard to the exploitation of corpus types, such as DIY corpora, specialized corpora and learner corpora, as well as relevant tools and applications. This gap is perceivable considering several challenges existing in corpus-assisted interpreting teaching.

To begin with, accessibility and corpus compilation present significant challenges, as spoken data are often more difficult to obtain and compile (Shlesinger 2009). Although current technological advancements in automatic speech recognition and other natural language processing tools have greatly facilitated the transcription of spoken data, the compilation of spoken corpora still faces tremendous hurdles in terms of data accessibility. This is partly due to confidentiality issues of the interpreted data and partly due to interpreters' reluctance to be examined and analyzed, whether for research or teaching purposes.

Moreover, the integration of corpus-based activities into interpreter training curricula requires careful planning and implementation. Unlike translation exercises, interpreting exercises may be more effective when students can actively engage in interactive and dynamic contexts. Bale (2015), for example, reported some reserved comments about the use of BACKBONE corpus in his interpreting classroom, with one student bluntly stating that "the interpreting exercises on the computer [felt] a little bit strange" due to a lack of interaction (p. 36). In addition to the inherent limitations of corpus-based activities, the functions of spoken corpora have also been underexplored in curriculum design. Trainers often focus narrowly on keyword lists or concordance lines without fully leveraging the broader potentials of corpus data. Braun and Chambers (2006) highlighted the utilization of spoken corpora for purposes beyond the creation of keyword lists and concordances, suggesting that interpreting trainers should design more comprehensive and engaging corpus-assisted exercises to motivate students and enhance their learning autonomy (cited from Bale 2015).

Besides the limited accessibility to spoken corpora and the under-exploitation of corpus functions, a more pressing problem is perhaps the lack of a systematic introduction of corpus methodologies to interpreting students. According to Mikhailov's (2022) survey, few universities have set up separate courses related to corpora for students majoring in translation and interpreting. More often than not, corpus studies are only "briefly introduced in theoretical courses in translation studies [...] or language technologies" (p. 20). The study also revealed a gap between corpus researchers and interpreting trainers, as most trainers are practitioners rather than researchers, and they themselves may lack a systematic understanding of corpus methodologies. This contributes to the peripheral role of corpora in actual interpreting training programs. This disconnection between theory and practice has long been a persistent issue in translation studies. To address this problem, urgent collaboration between translation and interpreting practitioners and researchers is necessary to better inform each other's practices and advance the field.

9.4.2 Suggestions and future directions

To address these challenges, several suggestions are proposed here. First and foremost, top priority should be given to a closer collaboration between corpus researchers and interpreting trainers to help bridge the gap between theory and practice. For example, interpreting trainers should stay updated with the latest developments in corpus-based interpreting studies and incorporate them into their classrooms, while researchers should familiarize themselves with the actual difficulties students encounter during their learning processes. Such collaboration can also lead to the development of specialized spoken corpora tailored to the needs of interpreting students, helping them address specific challenges they face in real-world interpreting scenarios.

Equally important is the incorporation of corpus linguistics courses into translation and interpreting curricula to equip both students and trainers with systematic methodologies and essential skills for effective data mining and analysis. These courses should cover both theoretical foundations and practical applications, enabling students to harness the power of corpora in their professional practice. Training programs for instructors can also ensure that they are well-equipped to integrate corpus-based activities into their teaching. This leads to the third proposal, which is designing more interactive and engaging corpus-based exercises that simulate real-life interpreting scenarios, such as role-playing and interactive simulations, as suggested by Dal Fovo (2018) and Xu (2018), among others.

In terms of future directions, corpus-assisted interpreter training should leverage recent technological developments, including virtual reality (VR) and augmented reality (AR) applications and the rise of generative artificial intelligence such as ChatGPT. Emerging technologies like VR and AR are beginning to find applications in interpreter training, as these immersive technologies can simulate real-life scenarios, providing students with a safe and controlled environment to practice their skills. Chan's studies (2022, 2023) vividly demonstrate the effectiveness of the Virtual Interpreting Practice (VIP) tool in enhancing both the interpreting and language competences of the examined students and improving their learning experience. The use of VR and AR in conjunction with corpus resources can further enhance the authenticity of training materials. Students can interact with virtual environments enriched with corpus-based data, enabling them to apply theoretical knowledge to practical situations. This combination of immersive technology and corpus tools represents a promising frontier in interpreter education.

Another direction concerns with the potential integration between corpus resources and ChatGPT, although their synergy has yet to be investigated thoroughly, either theoretically or empirically. Current discussions focus mainly on the effectiveness of ChatGPT in the translation classroom, particularly in comparison to other machine translation tools such as Google Translate or DeepL, or human translation. In the near future, interpreter training can leverage the vast data contained with electronic corpora, coupled with the adaptive learning capacities of AI, thus providing more dynamic and personalized learning experiences. For instance, ChatGPT can be used to identify patterns and trends within corpora, such as frequently used terminology or common errors, and assist in creating interactive exercises with the provided corpus data for both trainers and trainees. This synergy between corpus resources and AI technologies can enhance the comprehensiveness and efficiency of interpreter training, making it more adaptable to the needs of individual learners and more reflexive of the complexities of real-world interpreting tasks.

9.4   Conclusion

Electronic corpora play a vital role in translator and interpreter training, providing authentic data that enhances educational outcomes. This chapter reviewed the current state of research and practice, highlighting key findings and suggesting future directions for integrating corpora in translation and interpreting education. While significant progress has been made in translator training, interpreter training still faces challenges such as accessibility, corpus compilation and effective curricular integration.

Despite these challenges, the continued evolution of corpus methodologies, especially in the era of artificial intelligence, promises to further improve the training and education of translators and interpreters. As Bernardini et al. (2003) rightfully pointed out, the final goal of introducing corpus methodologies to translator and interpreter training is to make students "better language professionals in a working environment where computational facilities for processing texts have become the rule rather than the exception" (p. 2). Promoting these new working methods, as Mikhailov (2022) emphasized, requires overcoming inertia in both the industry and universities, thereby highlighting the crucial role of academic institutions in this advancement.

To bridge the gap between theory and practice, a closer collaboration between corpus researchers and interpreting trainers is essential. Incorporating systematic corpus linguistic course into translation and interpreting curricula, as well as leveraging emerging technologies such as VR and artificial intelligence, can significantly enhance the training process. The integration of these advanced technologies with corpus resources can provide more dynamic and personalized learning experiences, ultimately producing more proficient and adaptable language professionals.

References

Anthony, Lawrence. 2004. "AntConc: A Learner and Classroom Friendly, Multi-Platform Corpus Analysis Toolkit." In *Proceedings of IWLeL 2004: An Interactive Workshop on Language E-Learning* 7–13. Waseda, Tokyo.

Anthony, Lawrence. 2005. "AntConc: Design and Development of A Freeware Corpus Analysis Toolkit for the Technical Writing Classroom." In *IEEE International Professional Communication Conference Proceedings* 729–37. Limerick, Ireland.

Baker, Mona. 1993. "Corpus Linguistics and Translation Studies: Implications and Applications." In *Text and Technology: In Honour of John Sinclair*, edited by Mona Baker, Gill Francis, and Elena Tognini-Bonelli, 233–50. Amsterdam: John Benjamins

Baker, Mona. 1995. "Corpora in Translation Studies: An Overview and Some Suggestions for Future Research." *Target* 7(2): 223–43.

Bale, Richard. 2013. "Undergraduate Consecutive Interpreting and Lexical Knowledge. The Role of Spoken Corpora." *The Interpreter and Translator Trainer* 7(1): 27–50.

Bale, Richard. 2015. "An Evaluation of Spoken Corpus-based Resources in Undergraduate Interpreter Training." *International Journal of Applied Linguistics* 25(1): 23–45.

Barlow, Michael. 1995. *A Guide to ParaConc*. Athelstan: Houston.

Bendazzoli, Claudio. 2010. "The European Parliament as A Source of Material for Research into Simultaneous Interpreting: Advantages and Limitations." In *Translationswissenschaft–Stand und Perspektiven. Innsbrucker Ringvorlesungen zur Translationswissenschaft VI [Translation Studies, Status and Prospects. Innsbruck's Lecture Series on Translation Studies VI]* (Forum Translationswissenschaft, Band 12), edited by Lew N. Zybatow, 51–68. Bern: Peter Lang.

Bernardini, Silvia, Adriano Ferraresi, and Maja Miličević. 2016. "From EPIC to EPTIC— Exploring Simplification in Interpreting and Translation from an Intermodal Perspective." *Target* 28(1): 61–86.

Bernardini, Silvia, Dominic Stewart, and Federico Zanettin. 2003. "Corpora in Translator Education: An Introduction." In *Corpora in Translator Education*, edited by Federico Zanettin, Silvia Bernardini, and Dominic Stewart, 1–14. Beijing: Foreign Language Teaching and Research Press.

Biel, Łucja. 2011. "Professional Realism in the Legal Translation Classroom: Translation Competence and Translator Competence." *Meta* 56(1): 162–78.

Biel, Łucja. 2017. "Enhancing the Communicative Dimension of Legal Translation: Comparable Corpora in the Research-informed Classroom." *The Interpreter and Translator Trainer* 11(4): 316–36.

Bowker, Lynne. 2000. "Towards a Methodology for Exploiting Specialized Target Language Corpora as Translation Resources." *International Journal of Corpus Linguistics* 5(1): 17–52.

Bowker, Lynne, and Michael Barlow. 2008. "A Comparative Evaluation of Bilingual Concordancers and Translation Memory Systems." In *Topics in Language Resources for Translation and Localisation*, edited by Elia Yuste Rodrigo, 1–22. Amsterdam: John Benjamins.

Bowker, Lynne, and Peter Bennison. 2003. "Student Translation Archive: Design, Development and Application." In *Corpora in Translator Education*, edited by Federico Zanettin, Silvia Bernardini, and Dominic Stewart, 103–17. Manchester: St. Jerome.

Bowker, Lynne, and Jennifer Pearson. 2002. *Working with Specialized Language: A Practical Guide to Using Corpora.* London: Routledge.

Braun, Sabine, and Angela Chambers 2006. "Elektronische Korpora als Ressource für den Fremdsprachenunterricht." In *Praktische Handreichung für Fremdsprachenlehrer,* edited by Udo O. H. Jung, 330–37. Frankfurt am main: Peter Lang.

Brezina, Vaclav, Matthew Timperley, and Tony McEnery. 2018. #LancsBox v. 4.x. Lancaster, UK. Accessed January 8, 2025. https://www.research.lancs.ac.uk/portal/en/publications/lancsbox-v-4x(b03e99c8-7e4e-4915-927e-ef56a9999c5e).html.

Buendía-Castro, Miriam, and Clara Inés López-Rodríguez. 2013. "The Web for Corpus and the Web as Corpus in Translator Training." *New Voices in Translation Studies* 10(1): 54–71.

Chan, Venus. 2022. "Using a Virtual Reality Mobile Application for Interpreting Learning: Listening to the Students' Voice." *Interactive Learning Environment* 32(6): 1–14.

Chan, Venus. 2023. "Investigating the Impact of a Virtual Reality Mobile Application on Learners' Interpreting Competence." *Journal of Computer Assisted Learning* 39(4): 1242–58.

Dal Fovo, Eugenia. 2018. "The Use of Dialogue Interpreting Corpora in Healthcare Interpreter Training: Taking Stock." *The Interpreters' Newsletter* 23: 83–113.

EMT Expert Group. 2009. "Competences for Professional Translators, Experts in Multilingual and Multimedia Communication. Brussels, European Commission." Accessed January 8, 2025. https://www.scribd.com/document/356704637/emt-competences-translators-en-pdf.

Fantinuoli, Claudio. 2018a. "Interpreting and Technology: The Upcoming Technological Turn." In *Interpreting and Technology*, edited by Claudio Fantinuoli, 1–12. Berlin: Language Science Press.

Fantinuoli, Claudio. 2018b. "The Use of Comparable Corpora in Interpreting Practice and Training." *The Interpreters' Newsletter* 23: 133–49.

Gorjanc, Vojko. 2011. "Language Resources and Corpus-Driven Community Interpreter Training." In *Modelling the Field of Community Interpreting. Questions of Methodology in Research and Training,* edited by Claudia Kainz, Erich Prunc, and Rafael Schögler, 280–97. Münster: LIT Verlag.

Granger, Sylviane. 2013. "A Bird's-Eye View of Learner Corpus Research." In *Computer Learner Corpora, Second Language Acquisition and Foreign Language Teaching*, edited by Sylviane Granger, Joseph Hung, and Stephanie Petch-Tyson, 3–33. Amsterdam: John Benjamins.

Kelly, Dorothy. 2005. *A Handbook for Translator Trainers.* London: Routledge.

Kilgarriff, Adam, Pavel Rychly, Pavel Smrž, and David Tugwell. 2008. The Sketch Engine. In *Practical Lexicography: A Reader*, edited by Thierry Fontenelle, 297–306. Oxford: Oxford University Press.

Laursen, A. Lise, and Ismael A. Pellón. 2012. "Text Corpora in Translator Training: A Case Study of the Use of Comparable Corpora in Classroom Teaching." *The Interpreter and Translator Trainer* 6(1): 45–70.

Laviosa, Sara, and Gaetano Falco. 2023. "Corpora and Translator Education: Past, Present, and Future." In *Corpora and Translation Education: Advances and Challenges*, edited by Pan Jun and Sara Laviosa, 9–31. Singapore: Springer.

Liu, Kanglong. 2020. *Corpus-Assisted Translation Teaching.* Singapore: Springer.

Liu, Kanglong, Yanfang Su, and Dechao Li. 2023. "How Do Students Perform and Perceive Parallel Corpus Use in Translation Tasks? Evidence from an Experimental Study." In *Corpora and Translation Education: Advances and Challenges*, edited by Pan Jun and Sara Laviosa, 135–57. Singapore: Springer.

Manuel Jerez, Jesús. 2006. "Bringing Professional Reality into Conference Interpreter Training Through New Technologies and Action Research." PhD dissertation, University of Granada.

McEnery, Tony, and Andrew Hardie. 2012. *Corpus Linguistics: Method, Theory and Practice*. Cambridge: Cambridge University Press.

Mikhailov, Mikhail. 2022. "Text Corpora, Professional Translators and Translator Training." *The Interpreter and Translator Trainer* 16(2): 224–46.

Monzó, Esther. 2003. "Corpus-Based Teaching: The Use of Original and Translated Texts in the Training of Legal Translators." *Translation Journal* 7: 1–3.

Moratto, Riccardo. 2023. "Corpus-Assisted Translation Teaching: Issues and Challenges." *Translation & Interpreting* 15(1): 293–97.

Moropa, Koliswa. 2007. "Analysing the English-Xhosa Parallel Corpus of Technical Texts with Paraconc: A Case Study of Term Formation Processes." *Southern African Linguistics and Applied Language Studies* 25(2): 183–205.

Obrusnik, Adam. 2023. "Data Acquisition and Other Technical Challenges in Learner Corpora and Translation Learner Corpora." In *Corpora and Translation Education: Advances and Challenges*, edited by Pan Jun and Sara Laviosa, 161–69. Singapore: Springer.

PACTE Group. 2003. "Building a Translation Competence Model." In *Triangulating Translation: Perspectives in Process Oriented Research*, edited by Fabio Alves, 43–66. Amsterdam: John Benjamins.

Pan, Jun, Tak-Ming B. Wong and Honghua Wang. 2022. "Navigating Learner Data in Translator and Interpreter Training: Insights from the Chinese/English Translation and Interpreting Learner Corpus (CETILC)." *Babel* 68(2): 236–66.

Rodríguez-Inés, Patricia, and Amparo Hurtado Albir. 2012. "Assessing Competence in Using Electronic Corpora in Translator Training." In *Global Trends in Translator and Interpreter Training: Mediation and Culture*, edited by Séverine Hubscher-Davidson and Michal Borodo, 96–126. London: Bloomsbury Publishing.

Russo, Mariachiara. 2010. "Reflecting on Interpreting Practice: Graduation Theses Based on the European Parliament Interpreting Corpus (EPIC)." In *Translationswissenschaft: Stand und Perspektiven. Innsbrucker Ringvorlesungen zur Translationswissenschaft VI [Translation Studies: Status and Prospects. Innsbruck's Lecture Series on Translation Studies VI]* (Forum Translationswissenschaft 12), edited by Lew N. ZybatowBern, 35–50. Bern: Peter Lang.

Sandrelli, Annalisa. 2010. "Corpus-Based Interpreting Studies and Interpreter Training: A Modest Proposal." In *Translationswissenschaft: Stand und Perspektiven. Innsbrucker Ringvorlesungen zur Translationswissenschaft VI [Translation Studies: Status and Prospects. Innsbruck's Lecture Series on Translation Studies VI]* (Forum Translationswissenschaft 12), edited by Lew N. ZybatowBern, 69–90. Bern: Peter Lang.

Scott, Mike. 2008. WordSmith Tools, version 5. Liverpool, UK. Accessed January 8, 2025. https://www.lexically.net/wordsmith/downloads/#gsc.tab=0.

Shlesinger, Miriam. 2009. "Towards a Definition of Interpretese: An Intermodal, Corpus-Based Study." In *Efforts and Models in Interpreting and Translation Research: A Tribute to Daniel Gile*, edited by Gyde Hansen, Andrew Chesterman, and Heidrun Gerzymisch-Arbogast, 237–53. Amsterdam: John Benjamins.

Simonnæs, Ingrid, Jan Roald, and Beate Sandvei. 2023. "The Bergen Translation Corpus and Its Benefit for Training Purposes: A Case Study on Legal Texts." *The Interpreter and Translator Trainer* 17 (2): 282–300.

Spinzi, Cinzia. 2017. "Using Corpus Linguistics as a Research and Training Tool for Public Service Interpreting (PSI) in the Legal Sector." *The Interpreters' Newsletter* 22: 79–99.

Teich, Elke. 2003. *Cross-linguistic Variation in System and Text: A Methodology for the Investigation of Translations and Comparable Texts (Vol. 5)*. Berlin: De Gruyter Mouton.

Xu, Ran. 2018. "Corpus-Based Terminological Preparation for Simultaneous Interpreting." *Interpreting* 20(1): 29–58.

Zanettin, Federico. 1998. "Bilingual Comparable Corpora and the Training of Translators." *Meta* 43(4): 1–14.

Zanettin, Federico. 2009. "Corpus-Based Translation Activities for Language Learners." *The Interpreter and Translator Trainer* 3(2): 209–24.

10

ASSESSING INTERPRETING QUALITY USING LARGE LANGUAGE MODELS

An exploration

Masaru Yamada and Kayo Matsushita

10.1 Introduction

The assessment of simultaneous interpreting quality has long been a focal point in both academic research and interpreter training. Traditionally, this evaluation has been the purview of human experts, a process characterized by its intensive labor requirements and a notable lack of systematic methodology. Specifically, there has been an absence of a unified framework for quality assessment, in contrast to more recently established frameworks for written translation quality evaluation (Makinae et al. 2023; Matsushita and Yamada 2022; Yamada et al. 2023). This discrepancy highlights a foundational issue: the field of interpreting has yet to solidify a comprehensive framework for quality assessment, which is further complicated by a general lack of automation in assessment processes.

It is inaccurate to claim that no effort has been made to automate this evaluation. As of February 2024, Claudio Fantinuoli, a researcher and Chief Technology Officer (CTO) at KUDO, a US-based provider of a remote simultaneous interpreting platform, has already made public that the company has developed an automated system to evaluate interpreting quality. According to Fantinuoli (2024), this system assesses performance based on accuracy, fluency and latency. Moreover, some researchers have utilized natural language processing technologies to evaluate interpreting quality (Stewart et al. 2018; Zhang 2016).

However, some argue that the advent of Large Language Models (LLMs) such as ChatGPT has introduced a sense of obsolescence into these existing technological approaches. This evolution in technology not only impacts the domains of interpreting and translation assessment but also intertwines with a broader spectrum of

DOI: 10.4324/9781003597711-11

concern and anticipation regarding the potential for LLMs to replace or augment human interpreters and translators (Stojkovski 2023).

Considering these developments, this chapter explores the use of LLMs in conducting text-based assessments of the quality of human simultaneous interpreting. Given the rapid pace of technological advancement, this chapter does not present definitive experimental validation. Instead, it aims to probe the capabilities and potential of LLMs in performing interpreting tasks and, by extension, their utility in quality assessment. A key concept underpinning this exploration is prompt engineering, a methodology for crafting queries that leverage the interpretive and evaluative capabilities of LLMs. Specifically, we utilized ChatGPT-4. Through this lens, this chapter discusses how concepts central to the evaluation of interpreting quality can be effectively transformed into prompts, enabling a novel approach to assessment that combines traditional expertise with cutting-edge artificial intelligence (AI) technology.

10.2   Literature review

10.2.1   Review of existing frameworks for interpreting quality assessment

Scholars in interpreting studies have traditionally emphasized that interpreting quality is not a monolithic concept, but rather a dynamic interplay of various factors viewed differently by clients, interpreters, trainers and researchers (Collados Aís and García Becerra 2015). Error-based quality assessment has long existed (Barik 1971) and has been used in subsequent research (Gile 1999), but it has been challenged by more context-oriented perspectives (Pöchhacker 1994, 2001) and those focusing on the views held by users (Kurz 2001) or by the interpreters themselves (Zwischenberger 2010). However, there is no globally accepted standard framework for quality assessment and training in the interpreting industry (Yamada et al. 2023), and debates exist on whether a uniform standard should be applied (Pöchhacker 1994). This lack of consensus has led to variations in the perceptions of interpreting training instructors (Ahmed 2020).

Compared with translation, interpreting traditionally lacks text-based products for evaluation, making product-based assessments challenging. Recent advancements in technology such as automatic speech recognition and alignment tools have facilitated the construction of simultaneous interpreting corpora, enabling product-based quality evaluation. Nonetheless, a comprehensive framework for product-based quality assessment in simultaneous interpreting remains undeveloped, in contrast to translation quality assessment frameworks such as Multidimensional Quality Matrices (MQM),[1] which have seen broader adoption. Efforts are emerging to establish a product-based evaluation framework for simultaneous interpretating using MQM as a starting point (Makinae et al. 2023; Yamada et al. 2023).

10.2.2 Emergence of LLMs and their impact on translation and interpreting

The rise of LLMs, such as ChatGPT, has dramatically outperformed the accuracy of traditional neural machine translation (NMT) technologies. This leap forward not only demonstrates the rapid advancements in AI but also significantly influences the field of translation and interpreting. Research by Hendy et al. (2023) and similar studies underscores the important role of these models, particularly their superior ability to translate English into other languages, thereby affecting the industry at large (Moral and Mandell 2023). This includes notable improvements in the quality of English-to-Japanese translations (Sutanto et al. 2024). The significance of this development goes beyond merely enhancing the precision of translations; it represents a fundamental shift in the methodologies and practices of translation across various industries.

However, the contribution of LLMs to the translation sector is not limited to enhancing accuracy. Yamada (2023) delved into the transformative potential of incorporating the translation's purpose and the target audience's characteristics into the prompts used for generating translations with ChatGPT. By integrating insights from previous translation studies, industry practices and ISO standards, Yamada's research emphasizes the critical role of the preproduction phase in the translation workflow. This study demonstrates that by embedding appropriate prompts into ChatGPT, it is possible to achieve translations that are not only accurate but also adaptable to specific contexts, a capability that remains elusive in traditional MT systems.

10.2.3 The potential of automatic evaluation and LLM utilization in translation and interpreting

Product-based evaluations in interpreting have been challenging, leading to labor-intensive, subjective and impressionistic annotations, although some automatic and semi-automatic evaluation attempts have been made in natural language processing (Stewart et al. 2018; Zhang 2016). Automatic machine scoring is a transformative approach that uses algorithms to calculate metrics and indices of interpreting quality with minimal human intervention. While promising in terms of efficiency and consistency, this method requires further research to confirm its reliability and effectiveness. It spans a wide range of research in applied linguistics, computational linguistics and natural language processing with the aim to automate the interpreting quality assessment processes that are traditionally dependent on human judgment. Han and Lu (2021) explored the complex relationship between human rater scoring and automatic machine scoring in interpreting quality assessment, highlighting its significance in various social contexts and its implications for professional identity, accessibility and decision-making.

Conversely, the field of translation has seen significant advancements in automatic evaluation. Metrics such as Bilingual Evaluation Understudy (BLEU) (Papineni et al. 2002) and Translation Edit Rate (TER) (Snover et al. 2006) have become dominant, assessing how closely machine-generated translations resemble existing translations, which are considered golden standards. The advent of LLMs has significantly improved translation quality, surpassing the capabilities of metrics, such as BLEU scores, for quality assessment. This has led to the adoption of more rigorous evaluation methods such as those used in manual translation evaluations, including MQM. Recently, the application of LLMs for translation quality evaluation has been explored, with attempts to automate MQM annotations using tools such as ChatGPT (Fernandes et al. 2023; Kocmi and Federmann 2023). This indicates a growing expectation for the use of LLMs in translation quality evaluations, with potential implications for interpreting quality assessment as well.

10.3   Methodology

Against this backdrop, the aim of this study was to explore the extent to which LLMs can contribute to interpreting quality evaluations. Given the rapid evolution of technology, our focus is not to meticulously examine the accuracy of LLMs in replicating traditional quality assessment methods performed by humans, nor is it to attempt to replace automatic quality assessment tools used in natural language processing with LLMs. Instead, we seek to understand the extent to which current LLMs can effectively evaluate the interpreting quality.

However, the evaluative capacity of LLMs is not straightforward, especially considering that recent LLMs come equipped with conversational interfaces that allow for chat-based interactions. Therefore, the efficacy of an LLM is closely linked to the type of questions posed by users, essentially depending on the nature of the prompts used. Thus, this study first explains the fundamental concepts of prompt engineering and, following this approach, evaluates interpreting quality through specific examples related to prompt engineering.

10.3.1   Prompt engineering

Prompt engineering refers to the strategic formulation of input prompts to effectively communicate with and leverage the capabilities of LLMs, such as those used in conversational AI systems. This methodology is crucial for extracting the desired responses or behaviors from LLMs because their output is highly dependent on how queries are structured. In the context of LLMs, prompt engineering has become a critical tool for maximizing the utility of these models across various applications ranging from natural language processing tasks to complex problem-solving scenarios. This study delineates the essence of prompt engineering and evaluates its impact on the interpretative quality of LLM responses through specific examples, particularly focusing on three distinct types of prompts: zero-shot, few-shot and chain-of-thought (COT) prompts.[2]

10.3.1.1   Zero-shot prompts

Designed for scenarios in which the LLM is expected to generate a correct response without prior examples, zero-shot prompts test the model's ability to understand

and respond to queries based on preexisting knowledge and training without additional context or learning examples. They are pivotal for evaluating the base level of understanding of a model and its ability to generalize across different tasks without specific prior instructions.

10.3.1.2 Few-shot prompts

Involving providing the LLM with a small number of examples to guide its responses to new similar tasks, few-shot prompts leverage the model's learning capabilities by offering it a context or pattern to follow, thereby improving its accuracy and relevance to the responses. They are particularly useful in scenarios in which a certain level of tailored response is necessary; however, extensive training data are not available or feasible. Brown et al. (2020) introduced the concept of few-shot learning in the context of LLMs, providing a foundation for subsequent research on prompt engineering.

10.3.1.3 COT prompts

Designed to encourage the LLM to think aloud or follow a series of logical steps to arrive at an answer, COT prompts tackle complex problems that require reasoning, making it easier to understand the model's thought process and potentially improving the accuracy of its conclusions. Unlike few-shot prompts, which demonstrate exemplary cases to prompt the LLM to reason by analogy, COT prompts explicitly outline the reasoning behind the conclusion. For instance, when evaluating an interpreted output, a COT prompt might instruct, "consider the interpretation as good if it does not omit any important information from the original text", thereby guiding the model to reveal the thought process behind its judgment. Wei et al. (2022) presented a novel approach that encourages models to exhibit reasoning processes and offers insights into the potential for more complex problem-solving capabilities in LLMs.

In summary, prompt engineering encompasses a range of strategies aimed at enhancing the interaction with and output from LLMs. By understanding and applying the principles of zero-shot, few-shot and COT prompts, researchers and practitioners can harness the power of LLMs for various tasks more effectively. These strategies highlight the importance of the human role in guiding AI interactions and underscore the adaptability and potential of LLMs when faced with a wide array of challenges.

10.3.2 Exploration of LLM capabilities for interpreting quality evaluation

In this study, we employ a methodology that utilizes three distinct prompt types for interpreting quality assessments. Our analysis is grounded in the Japan National

Press Club (JNPC) Interpreting Corpus (Matsushita et al. 2020), a resource that facilitates research and education in Translation and Interpreting Studies. This English-Japanese bilingual corpus, which aggregates transcribed speeches (source text) and their subsequent interpretations (target text) from JNPC-hosted press conferences, is publicly accessible online.[3] For this study, we use excerpts from a 2015 press conference attended by Congressman Paul Ryan and a bipartisan US House delegation to assess the performance and precision of quality assessment by ChatGPT-4.[4] By limiting the scope of our analysis to selected segments from this corpus, we aim to showcase the potential of LLMs in interpreting quality assessments, emphasizing their applicability and effectiveness in real-world scenarios.

In conducting this verification of ChatGPT-4's capabilities, the study opted not to utilize the API; instead, it employed a chat interface available to regular paid users. This approach was chosen to simulate the user experience under standard conditions, thereby enhancing future adaptability. Furthermore, the temperature setting in ChatGPT-4, which adjusts the balance between predictability and creativity in its responses, remained unadjusted to preserve the default interaction dynamics. It is pertinent to note that the findings presented herein reflect the state of the technology as of the beginning of February 2024, providing a timely snapshot of its performance and user interaction nuances.

10.4   LLM prompt results on interpreting quality

10.4.1   Zero-shot prompt

First, we assessed the interpreting quality using a zero-shot prompt. In the context of a zero-shot prompt, evaluation was conducted without the benefit of prior examples or preexisting knowledge. The prompts devised by the authors are as follows:

> The following [source text] and [target text] are a corpus of simultaneous interpreting from English to Japanese. Please evaluate the quality of the [target text] against the [source text], bearing in mind that it was interpreted simultaneously.

Following the above prompt, texts from the JNPC Corpus, which included transcriptions of both speakers' renderings and interpreters' translations, were provided as part of the prompt in the form of [source text] and [target text] for analysis. [Back translation] was provided by the authors to facilitate understanding in this study.

[source text] We first want to say on behalf of ourselves in our country. We want to send our sympathies and condolences to the families of the victims of the ISIL a terrorist act. We were very impressed with the way the Japanese government handled the situation and we want to offer condolences to the families of the victims.

Second we are encouraged of and we're inspired by Prime Minister Abe on economic plans, especially his third arrow. We are eager to see the third arrow reforms implemented and were eager to see them occur so that Japan's economy can grow and strengthen because we believe of strengthening our two economies strengthening our alliances is vital and important for our people.

[target text] 冒頭に申し上げたいんですけれども私どもまた私どもの国々は心より哀悼の意をまず表させていただきます。ISILのテロの被害に遭った方々にまずお悔やみ申し上げます。すばらしく日本はこの件をハンドルなさったと思いまして、ご家族に哀悼の意を表します。これが第一点。第二点なんですけど非常に私触発受けているんですね。安倍総理の経済計画について。特に三本目の矢についてです。というのはわれわれとしてもぜひこの三本の目の矢がちゃんと実行できるようにそして成果を出していただきたいと思っています。さらに日本の経済に成長していただきたいからです。というのは日米の経済が伸びるということを我々の日米同盟関係が強化されるということはわれわれの国民にとって欠かせない重要なことだからです。

[back translation] Let us begin by saying that we and our countries extend our heartfelt condolences to the victims of the ISIL terrorist attacks. We wish to extend our deepest sympathies to those affected by the ISIL terrorist attacks. We think Japan has handled this wonderfully, and we offer our condolences to the family. This is the first point. Secondly, we are very much inspired, by Prime Minister Abe's economic plan. In particular, we would like to talk about the third arrow. We are very much looking forward to the implementation of these three arrows and the results that they will produce. This is because we would like to see Japan's economy grow even more. The growth of the Japanese and US economies, and the strengthening of the Japan-US alliance, are of vital importance to the people of Japan.

The outcomes derived from the given prompts are as follows: It is worth noting that even with a zero-shot prompt (i.e., with no context provided), detailed evaluation results using quality criteria common in human-based assessments were generated.

ChatGPT's response:

Evaluating the quality of simultaneous interpreting involves considering several factors, including accuracy, completeness, coherence, and the ability to convey the original tone and nuances. Here's an assessment based on these criteria:

Accuracy and Completeness

The [target text] captures the essence and key points of the [source text] quite well. It conveys the message of sympathy and condolences for the victims of the ISIL terrorist act and acknowledges the Japanese government's handling of the situation. It also accurately reflects the encouragement and inspiration drawn from Prime Minister Abe's economic plans, especially mentioning the "third arrow" of his strategy.

However, there are some minor omissions and simplifications, which are common in simultaneous interpreting due to the need to keep up with the speaker in real

time. For example, the phrase "on behalf of ourselves and our country" is simpli-
fied to "私どもまた私どもの国々は" (we and our countries), which streamlines
the message while retaining its essence.

Coherence

The [target text] maintains coherence and logical flow, successfully dividing the
content into two main points as in the [source text]. The interpreter uses connec-
tors and transitional phrases appropriately ("冒頭に申し上げたいんですけれ
ども" at the beginning, "これが第一点" for the first point, "第二点なんですけ
ど" for the second point), which helps in structuring the interpretation cohesively.

Tone and nuances

The interpreter manages to convey the tone of sympathy, respect, and encourage-
ment found in the [source text]. Phrases like "心より哀悼の意を表させていた
だきます" (express our deepest sympathies) and "非常に私触発受けているん
ですね" (very much inspired) align well with the emotional and respectful tone of
the original message.

Overall evaluation

The [target text] is a high-quality simultaneous interpretation of the [source text].
It achieves a good balance between fidelity to the original, coherence, and fluency,
which are critical in simultaneous interpreting. Despite the inherent challenges
of interpreting in real time, such as the need to condense or paraphrase for effi-
ciency, the interpreter successfully conveys the main messages and sentiments of
the speaker. This indicates a high level of skill and professionalism in interpreting.

As observed, the assessment by ChatGPT displays a satisfactory level of precision despite the prompt's lack of detailed instructions. Qualitative analysis criteria,
such as completeness and coherence, are aligned with those emphasized in prior
research (Zwischenberger 2010). The overall evaluation presented the findings in
an organized and comprehensible manner. However, the consistently polite and
positive tone typical of current-day LLMs, coupled with the absence of construc-
tive criticism, renders the assessment somewhat superficial compared to that of
human evaluators or instructors.

Further scrutiny revealed that ChatGPT's evaluations lacked detailed explanations to support its conclusions. For instance, the interpreter's translation of "on
behalf of ourselves and our country" into "私どもまた私どもの国々は" (we and
our countries) is intended to mirror the original text closely, striving to encapsulate
"ourselves" and "our country" with the reiterated phrase. However, this attempt
misses some elements, such as the exclusion of "on behalf of" and the alteration
from "country" to "countries". ChatGPT characterizes these discrepancies merely

as "minor omissions and simplifications". While this explanation might be applicable to the exclusion of "on behalf of", the use of the plural "countries" instead of the singular "country" should be regarded as a more significant error in this context given that Minister Bishop is only representing her country, Australia. Furthermore, it describes the translation's effect as "streamlining the message while preserving its core intent", yet the actual repetition ("私ども") introduces a redundancy that detracts from the fluidity of the interpretation.

Considering the nature of the initial inquiry as a zero-shot prompt, however, lacking detailed information or a contextual background, the critique formulated by ChatGPT cannot be dismissed as ineffective. On the contrary, from the standpoint of authors proficient in both English and Japanese, ChatGPT's evaluation resonates with a reasonable level of correctness. It showcases its adeptness in providing a valid assessment of the translation's overall quality, underlining the model's capability to generate meaningful insights, even when faced with limited guidance.

10.4.2   Zero-shot prompt with errors in the target text

As discussed in the previous section, the authenticity and precision of ChatGPT's ability to assess interpreting quality are subject to scrutiny. Consequently, we introduce an additional process to validate this evaluation. The original source texts remained the same, yet the target texts were deliberately altered to incorporate errors, such as omissions and mistranslations. This approach aims to test the capacity of ChatGPT to detect and precisely evaluate translation inaccuracies. The modifications are indicated by strike-throughs for omissions and boxes for specific errors. Subsequent analysis explored how ChatGPT responded to and assessed these intentionally flawed translations. Using the same prompts as above, the target text was replaced with the following:

[target text] 冒頭に申し上げたいんですけれども私どもまた私どもの国々は心より哀悼の意をまず表させていただきます。ISILのテロの被害に遭った方々にまずお悔やみ申し上げます。すばらしく日本はこの件をハンドルなさったと思いました。ご家族に幸せです。これが第一点。第二点なんですけど非常に私触発受けているんですね。安倍総理の経済計画について。特に三本目の矢についてです。というのはわれわれとしてもぜひこの三本の目の矢がちゃんと実行できるようにそして成果を出していただきたいと思ってます。さらに日本の経済に成長していただきたいからです。というのは日米の経済が伸びるということを我々の日米同盟関係が強化されるということはわれわれの国民にとって欠かせない重要なことだからです。

[back translation] Let us begin by saying that we and our countries extend our heartfelt condolences to the victims of the ISIL terrorist attacks. We wish to extend our deepest sympathies to those affected by the ISIL terrorist attacks. We think Japan has handled this wonderfully, and we offer our happiness *to the*

family. This is the first point. Secondly, we are very much inspired, by Prime Minister Abe's <u>economic plan. In particular,</u> we would like to talk about the third arrow. We are very much looking forward to the implementation of these three arrows and the results that they will produce. This is because we would like to see Japan's economy grow even more. The growth of the Japanese and US economies, and the strengthening of the Japan-US alliance, are of vital importance to the people of Japan.

The following section presents ChatGPT's responses. As can be understood from reading the entire text, the focus is on whether ChatGPT appropriately identifies omissions and semantic errors in the target text. Moreover, it addresses whether the evaluation of the interpreting quality considers these errors. The format for the quality analysis is largely the same as the initial response to the prompt; however, the sections related to errors in the target text, which the authors had embedded, are mentioned. In the following response, the parts related to these sections are underlined.

ChatGPT's response[5] (emphasis added)

To evaluate the quality of the [target text] (Japanese) against the [source text] (English) in the context of simultaneous interpretation, we'll consider several key factors: accuracy, completeness, coherence, and the adaptation of cultural nuances. Simultaneous interpreting is challenging, requiring interpreters to instantly convert the speaker's messages into another language while preserving the original meaning, tone, and intent as closely as possible.

Accuracy

The [target text] accurately captures the essence of the [source text] in terms of expressing sympathies and condolences to the families of the victims of the ISIL terrorist act, the commendation of the Japanese government's handling of the situation, and the mention of Prime Minister Abe's economic plans, specifically the third arrow. However, <u>there are some inaccuracies and omissions in the translation</u>.

Completeness

The [target text] omits several details from the [source text]:

- *<u>The explicit mention of ISIL as a terrorist act is not directly translated.</u>*
- *<u>The repeated expression of offering condolences to the families of the victims is simplified and not fully conveyed in the [target text].</u>*
- *<u>The [target text] does not fully express the encouragement and inspiration drawn from Prime Minister Abe's economic plans, nor does it detail the eagerness to see the third arrow reforms implemented and their effects on strengthening the economies and alliances.</u>*

Coherence

The [target text] maintains coherence throughout, with a logical flow that mirrors the original speech. However, <u>the omission of some details may affect the listener's full understanding of the depth of sentiment and the specific elements of encouragement regarding economic plans</u>.

Cultural and contextual adaptation

Simultaneous interpretation often requires adapting cultural references and nuances to the target language audience. The [target text] does this well by using respectful language appropriate for expressing condolences and admiration. However, <u>the brief and somewhat vague mention of "ご家族に幸せです" (which translates to "happiness to the families") is an inaccurate and potentially confusing rendering of the intent to offer condolences,</u> indicating a possible misinterpretation or overly concise translation effort.

Overall evaluation

The [target text] provides a generally faithful interpretation of the [source text], capturing the main messages and sentiments. However,<u> it falls short in terms of completeness and specific accuracy, notably in conveying the full scope of the condolences, the direct reference to ISIL, and the detailed expression of support for Prime Minister Abe's economic strategies. These omissions and inaccuracies could lead to a less informed or differently nuanced understanding by the Japanese-speaking audience.</u>

In simultaneous interpreting, the balance between speed and detail is crucial. While the interpreter successfully communicated the key points and maintained coherence, the translation would benefit from greater precision and completeness to ensure that the audience fully grasps the depth and specifics of the original message.

The analysis indicated that ChatGPT-4, even without specialized adjustments or targeted pretraining, can detect mistranslations and missing elements in translations. This suggests that the model is effective in handling linguistic evaluations to a certain extent. This is observed through its ability to point out inaccuracies, including not directly translating references to ISIL and the inappropriateness of translating "ご家族に幸せです" as "happiness to the families" instead of offering condolences. These examples demonstrate the model's utility in assessing translation quality, which is noteworthy given that this is achieved through a zero-shot prompt.

Moreover, when considering the overall evaluation of interpreting, especially in contexts such as educational settings, where the comparison between interpretations with numerous omissions versus those accurately rendered is critical, the findings suggest that even without explicit training, ChatGPT-4's zero-shot prompts can differentiate between the two. This ability to discern quality differences in translations without specific preparation highlights the potential of AI in supporting translation and interpreting assessment tasks.

10.4.3 Few-shot prompt

This section explores the concept of few-shot prompting and contrasts it with zero-shot prompting by focusing on providing ChatGPT-4 with a small number of examples to guide its responses to new, similar tasks. This method was applied to evaluate the interpreting segments as either *good* or *bad*. A few examples are provided in advance as prompts, serving as benchmarks for the LLM to assess new translations. Additionally, a detailed evaluation was provided. Few-shot prompting comes from this preparatory step of supplying examples of good and bad translations.

In the few-shot prompt setup, examples of source text 1 were shown alongside target texts 1 and 1b, labeled to indicate which was considered *good* (// *good interpreting*) and which was *bad* (// *bad interpreting*). The annotations were made by the authors. The LLM was trained using these examples. It then evaluates whether the new targets (texts 4 to 6) are *good* or *bad* based on the learned standards. Importantly, the examples used here are consistent with those in the zero-shot prompt, blending the good and bad examples to discern the LLM.

In other words, the evaluation of the quality of target texts 4–6 has been previously established. We know that target texts 4 and 5 are *good*, albeit with noticeable omissions or inaccuracies in meaning, whereas target text 6, with significant omissions, is classified as a *bad* rendition according to the assessment derived from ChatGPT's zero-shot prompting response in the preceding section. Consequently, the focus of this investigation was to ascertain whether ChatGPT-4, when furnished with a few-shot prompt sample, can accurately discern these quality distinctions. The prompt administration for this purpose was delineated subsequently.

The following text pairs consist of source material in English and their corresponding translations into Japanese through simultaneous interpreting. Given that these translations are products of simultaneous interpreting, assess if the translations accurately reflect the source material. Examples of both effective and ineffective interpreting are provided for comparison. For instance, [source text 1] is accurately translated in [target text 1], demonstrating good interpreting, whereas [target text1b] represents poor interpreting. Based on these standards, evaluate whether translations 4 through 6 are examples of good or bad interpreting, providing explanations for your judgments.

[source text 1] We first want to say on behalf of ourselves in our country. We want to send our sympathies and condolences to the families of the victims of the ISIL a terrorist act.

[target text 1] 冒頭に申し上げたいんですけれども私どもまた私どもの国々は心より哀悼の意をまず表させていただきます。ISIL のテロの被害に遭った方々にまずお悔やみ申し上げます。// *good interpreting*

[target text1b] 冒頭に申し上げたいんですけれども私どもまた私どもの国々は心より哀悼の意をまず表させていただきます。// *bad interpreting*

[source text 2] We were very impressed with the way the Japanese government handled the situation

[target text 2] すばらしく日本はこの件をハンドルなさったと思いまして // *good interpreting*

[target text 2b] すばらしく日本はこの件をハンドルなさったと思いました // *good interpreting*

[source text 3] and we want to offer condolences to the families of the victims.

[target text 3] ご家族に哀悼の意を表します。// *good interpreting*

[target text 3b] ご家族に幸せです。// *bad interpreting*

[source text 4] Second we are encouraged of and we're inspired by Prime Minister Abe on economic plans, especially his third arrow.

[target text 4] これが第一点。第二点なんですけど非常に私触発受けているんですね。安倍総理の経済計画について。特に三本目の矢についてです。

[source text 5] We are eager to see the third arrow reforms implemented and were eager to see them occur

[target text 5] というのはわれわれとしてもぜひこの三本の目の矢がちゃんと実行できるようにそして成果を出していただきたいと思ってます。

[source text 6] so that Japan's economy can grow and strengthen because we believe of strengthening our two economies strengthening our alliances is vital and important for our people.

[target text 6] さらに日本の経済に成長していただきたいからです。

To encapsulate this, ChatGPT-4 responses were aligned with anticipated answers. It rated target texts 4 and 5 as *good* and target 6 as *bad*. Due to space limitations, only a part of ChatGPT-4's analysis is included here, specifically the sections relevant to explaining the answers. A complete analysis can be accessed online.[6]

Despite the limited number of samples provided for verification, the precision of the few-shot prompt evaluation improved significantly compared to that of the zero-shot prompt. The authors included in the prompt a request for an explanation of the judgments, thus obtaining from ChatGPT not only the decisions of *good* or *bad* but also the reasons behind these judgments.

ChatGPT's response (few-shot prompt)[7]

...

Evaluation of Translations 4 through 6
Translation 4 Evaluation

...

- Judgment: This is generally good interpreting, as it conveys the main idea of inspiration from Prime Minister Abe's economic policies, especially the third arrow. However, the omission of "encouraged" slightly diminishes the full sentiment expressed in the source text. It still effectively communicates the key aspects of the message, making it a minor deviation rather than a critical error.

Translation 5 Evaluation

...

- Judgment: This is good interpreting. The translation accurately captures the eagerness and anticipation for the implementation and success of the third arrow reforms, aligning well with the source material's intent.

Translation 6 Evaluation

...

- Judgment: This is a case of incomplete interpreting. While the target text successfully conveys the hope for Japan's economic growth, it significantly omits the broader geopolitical context and rationale behind this wish, specifically the strengthening of alliances and its importance. This omission detracts from the full understanding of the source text's intent, making it less effective.

In summary, translations 4 and 5 are examples of good interpreting, effectively conveying the main ideas and sentiments of their respective source texts, despite minor omissions in translation 4. Translation 6, however, represents a less effective interpreting effort due to its significant omission of key elements of the source text's message.

As seen in this example, few-shot prompts can be considered for use in educational contexts, even if there is a small corpus of exemplary interpretations. For instance, outputs by skilled professional interpreters can be set as the benchmark for "good interpreting", and evaluating each student's interpreting performance data against it could assist in the efficient and fair assessment of the performance of multiple students.

10.4.4 COT prompt

In this section, we employ the COT prompting technique to execute the proposition-based evaluation method described below. The prompts examined thus far were primarily intended to introduce evaluation methods that could be conducted without a reference (e.g., a rubric or list of quality criteria). In other words, the purpose was to investigate the extent of preliminary knowledge LLMs have concerning the act of interpreting and interpreting quality. Thus far, we have seen that even with zero-shot and few-shot prompts, LLMs can perform reasonably reliable evaluations. COT prompts serve as a strategy to explicitly define the criteria for *good interpreting* and to make quality judgments. Specifically, it involves describing the evaluation criteria and thought process, such as identifying important information, checking for omissions in the interpretation, calculating the rate of information agreement between the original speech and interpreting and then providing this description to the LLM for its assessment.

Before conducting this study, we explored the possibility of partially automating the interpreting quality assessment. We examined methodologies for assessing interpreter outputs using measurable and objective metrics such as correct terminology and the frequency/duration of silent pauses (Matsushita and Yamada 2023), partially employing existing frameworks for evaluating translation quality, such as MQM. However, translation quality frameworks cannot be directly applied to the evaluation of interpreting quality without significant modifications or customization. This is particularly pertinent to simultaneous interpreting, where it may be unavoidable, due to time constraints, to ensure the transmission of essential information without omissions, whereas less critical information may need to be selectively omitted.

Professional interpreters have anecdotally mentioned this concept. For example, Komatsu (2005), a leading English-Japanese simultaneous interpreter, stated that it is crucial to "distinguish the trunk from the branches and leaves", or, in other words, to discern between important and less important information while interpreting. Someya (2017) agreed fundamentally with Komatsu's (2005) point of view, proposing that this discernment of information importance can be applied to notetaking in consecutive interpreting, thereby outlining the skeleton of the original discourse (speech) based on the trunk and branches concept.

Similar views can be seen from the perspective of interpreting quality assessments. Prior research suggests that unit-based accuracy analysis is a common method for assessing sense consistency in simultaneous interpreting (Gieshoff and Albl-Mikasa 2022). The basic procedure involves dividing the source speech into propositions or small-meaning units and checking whether each unit is present in the target speech. The idea is to assess quality based on whether these propositions are translated.

This study did not aim to discuss the validity or reliability of the evaluation method. However, we can operationally assume that the points presented above are valid, allowing us to evaluate semi-automatically how accurately important information or propositions are translated in simultaneous interpreting. The objective of this section is to verify whether this can be achieved using LLMs. To achieve this, we used a COT prompt to provide the necessary instructions to the LLM. The prompt shown below is designed to command the LLM in the form of a COT to evaluate the ratio of translated information and whether the essential information (the "trunk") has been appropriately translated. The ultimate goal is to develop a system that can quantify the amount of translated information. We have made some exploratory steps, as shown in the following example.

A prompt was created by describing the five steps necessary for evaluation and instructing the LLM to execute them sequentially. The first step involves rewriting the original and translated texts into a propositional form with examples provided. This incorporates the elements of few-shot prompts as discussed earlier. In step 2, the "trunk" within the transformed propositional form is identified. Step 3 involves calculating the amount of information using the simple calculation method outlined in the prompt. Step 4 examines the degree of match in the amount of information between the original and translated texts and presents it as a numerical ratio. The final step involves creating a prompt that asks for a comprehensive evaluation of interpreting quality, considering the amount of information translated and whether the "trunk" has been captured.

Conventionally, achieving this with natural language processing technology requires complex programming skills. Furthermore, defining high-level abstract instructions, such as "identify the trunk", would have been nearly impossible. However, this can be achieved using proposition-based prompts, such as those below, and the outcome can be published online.[8]

10.4.4.1 COT prompt

We are assessing the quality of translations produced through simultaneous interpreting from English to Japanese. Below, we provide the [source text] and the [translated text].

Your task is to compare the [source text] and [target text]. It is widely recognized that simultaneous interpreting may not capture all the details of the source text, leading to omissions, redundancies, and errors. Our objective is to evaluate how effectively the translation conveys the core information – or the "trunk" information – of the source text. Please analyze the provided texts and follow these steps:

1 *Conversion to Propositional Formulas:*

   – *Convert the [source text] and [target text] into propositional formulas. Here's an example focusing on the predicate "eat":*
   Ken may eat a piece of cake
   : M A Y (: E A T (, p i e c e _ o f _ c a k e)) M : M A Y (P:EAT(Ken,piece_of_cake)).

2 *Identify Key Information Elements:*

   – *Extract the key information elements from the source text. Distinguish between the crucial "trunk" information and the supplementary "branches and leaves". This step focuses on identifying the "trunk" from the set of propositions in the source text.*

3 *Calculate the Amount of Information:*

   – *Based on steps 1 and 2, calculate the information content. Typically, each predicate is assigned an information value of 1. However, propositions identified as "trunk" information are valued at 2, indicating double the importance. Perform this calculation for both the source and the translated texts.*

4 *Calculate the Percentage of Information Agreement:*

 – *Determine the match rate of variables between the source and target propo-sitional formulas. Calculate the information agreement percentage by divid-ing the information content of the translated text by that of the source text. If the "trunk" information from the source text is missing in the target text, a lower information agreement ratio is expected.*

5 *Evaluate and Explain Interpreting Quality:*

 – *Assess the quality of the interpreting by examining the differences in informa-tion agreement rates, the variations in propositional formulas, and how well the "trunk" and "branches" correspond between the source and target texts.*

 --

The original and translated texts used for the evaluation were the same as those used for the zero-shot prompts. The results are shown below.

ChatGPT's response

Let's proceed with the analysis according to the outlined steps.

1 *Conversion to Propositional Formulas*
 Source Text Propositions:

 1 *WANT(speaker, SEND(sympathies, families_of_victims(ISIL_terrorist_act))).*
 2 *IMPRESSED(speaker, HANDLE(Japanese_government, situation)).*
 3 *WANT(speaker, OFFER(condolences, families_of_victims)).*
 4 *ENCOURAGED(speaker, ECONOMIC_PLANS(Abe, third_arrow)).*
 5 *EAGER(speaker, SEE_IMPLEMENT(third_arrow_reforms)).*
 6. *BELIEVE(speaker, VITAL(STRENGTHEN(economies), STRENGTHEN (alliances), FOR_PEOPLE)).*

 Target Text Propositions:

 1 *WANT(speaker, EXPRESS(sympathies, families_of_victims(ISIL_terrorist_ act))).*
 2 *THINK(speaker, HANDLE_WELL(Japanese_government, situation)).*
 3 *EXPRESS(speaker, condolences, families_of_victims).*
 4 *INSPIRED(speaker, ECONOMIC_PLANS(Abe, third_arrow)).*
 5 *WANT(speaker, SUCCESS(third_arrow_reforms)).*
 6 *BELIEVE(speaker, IMPORTANT(STRENGTHEN(economies), STRENGTHEN (alliances), FOR_PEOPLE)).*

2 *Identify Key Information Elements*
 Source Text "Trunk" Information:

 – *Condolences and sympathies to ISIL victims' families.*
 – *Impression of the Japanese government's handling of the situation.*

- *Encouragement and inspiration from Abe's economic plans (third arrow).*
- *The belief in the importance of strengthening economies and alliances.*

Target Text "Trunk" Information:

- *The expression of sympathies to ISIL victims' families.*
- *The positive view of the Japanese government's handling.*
- *Inspiration from Abe's economic plans, particularly the third arrow.*
- *The belief in the importance of economic and alliance strengthening.*

3 *Calculate the Amount of Information*
   Source Text:

- *Total propositions: 6*
- *"Trunk" propositions: 4 (Each valued at 2 for a total of 8)*
- *Total information value: 8 (since the other propositions support these main ideas)*

Target Text:

- *Total propositions: 6*
- *"Trunk" propositions: 4 (Each valued at 2 for a total of 8)*
- *Total information value: 8*

4 *Calculate the Percentage of Information Agreement*
   Both the source and target texts have an information value of 8, based on the "trunk" information.
   Information Agreement Percentage = (Target Information Value/Source Information Value) 100% = (8/8) 100% = 100%

5 *Evaluate and Explain Interpreting Quality*
   The interpreting quality is high, as evidenced by the 100% information agreement percentage. Both the source and target texts focus on the essential elements, such as expressing sympathies and condolences, acknowledging the Japanese government's response, being inspired by economic plans, and the importance of strengthening economic and alliance ties. The propositional analysis indicates that the core messages and sentiments were effectively preserved in the translation, demonstrating a successful conveyance of "trunk" information. Minor nuances and stylistic differences do not detract from the overall fidelity and completeness of the interpreted message.

As illustrated by the details in the previous response from ChatGPT, the analysis was conducted by executing five steps as a COT: (1) convert to propositional formulae; (2) identify key information elements; (3) calculate the amount of information; (4) calculate the percentage of information agreement; and (5) evaluate and explain interpreting quality. From this process, it can be determined that the target text of a good interpretation example achieved an information agreement percentage of 100.

The results clearly demonstrate the feasibility of conducting a detailed proposition-based quality assessment using LLMs. Interestingly, in the calculation of the ratio of translated information, instructions were given to assign double weighting to the propositions that form the "trunk" (step 3), and this was accurately executed. Although space constraints prevented us from including it here, it appears that the LLM was also able to correctly calculate the information ratio for "poor interpreting examples". Although further verification is necessary, the successful implementation of complex evaluation instructions using COT prompts with LLMs seems promising.

10.5 Discussion and conclusion

This study explored the application of three types of LLM prompts – zero-shot, few-shot and COT – to assess the quality of simultaneous interpreting. These findings suggest that by leveraging the logical processing capabilities of LLMs, a precise quality assessment of interpreting performance is possible and valuable insights can be obtained even without explicit training. The advantages of using LLM prompts are significant, including their potential for large-scale automated assessments. This could reduce the reliance on labor-intensive human evaluation and provide relatively consistent assessments across multiple dimensions of interpreting quality.

The evaluation method for interpreters using general prompts in LLMs, as proposed in this study, shows the potential for assessing complex translations performed by human interpreters. This technique can also be extended to other educational contexts. Interpreter training programs worldwide encounter challenges in providing courses for less common language pairs because of limited enrollment. This method allows the provision of specific feedback on language pairs, even in the absence of instructors. Furthermore, it can aid instructors in mitigating their biases by referencing outputs from LLMs. As demonstrated by the few-shot prompt examples, a small corpus of high-quality interpretations can serve as an assessment benchmark. This method also offers practical self-learning opportunities for interpreters and students, combining high utility and minimal cost. In addition, a detailed proposition-based analysis, as exemplified by the COT prompt, can be employed for evaluative purposes in interpreting studies.

Future research should focus on ways to integrate LLMs into interpreter training and assessment frameworks and investigate how these technologies can complement traditional methods. Enhancing the sensitivity of LLMs to cultural and contextual nuances is crucial, potentially through the development of more sophisticated prompt engineering techniques. However, it is essential to recognize that these are text-based assessments of interpreting quality and have inherent limitations. The challenge lies in incorporating elements and characteristics that can only be confirmed through audio data, such as the interpreter's tone of voice, use of prosody and other delivery-related factors, into the evaluation process.

Additionally, the reliance on prompt engineering highlights the need for specialized knowledge to craft effective prompts, which could limit broader applicability without significant training or expertise. Addressing these limitations in future research could further refine the use of LLMs in interpreting quality assessments, ensuring a more comprehensive and detailed evaluation process.

Notes

1 MQM (Multidimensional Quality Metrics). Retrieved from https://themqm.org/introduction-to-tqe/an-overview/.
2 Prompting Guide. Retrieved from https://www.promptingguide.ai/.
3 JNPC Interpreting Corpus. Retrieved from https://www.gsk.or.jp/catalog/gsk2020-a/.
4 OpenAI. Retrieved from https://openai.com/
5 ChatGPT's response (zero-shot prompt) is available at https://chat.openai.com/share/49fe8083-1874-4ff1-9417-7a08cbf3b67a.
6 https://chat.openai.com/share/b36f8a03-daa4-4818-8b7a-660f25b1f8f6.
7 ChatGPT's response (few-shot prompt) is available at https://chat.openai.com/share/6949767b-f8e6-47ae-9806-9a423b3e375a.
8 COT prompt (Proposition-Based Analysis for Simultaneous Interpreting Quality Assessment) is available at: https://chat.openai.com/g/g-XVPM4mZWt-simultaneous-interpreting-quality-assessment.

References

Ahmed, Safaa. 2020. "Quality Assessment of Simultaneous Interpreting: Teaching and Learning Perspective to English and Arabic Renditions." *Manchester Journal of Artificial Intelligence & Applied Science (MJAIAS)* 1(1): 55–61.

Barik, Henri. 1971. "A Description of Various Types of Omissions, Additions and Errors of Translation Encountered in Simultaneous Interpretation." *Meta* 16(4): 199–210. https://doi.org/10.7202/001972ar.

Brown, Tom B., Benjamin Mann, Nick Ryder, Melanie Subbiah, Jared Kaplan, Prafulla Dhariwal, and Arvind Neelakantan et al. 2020. "Language Models Are Few-Shot Learners." ArXiv. July 22, 2020. https://doi.org/10.48550/arXiv.2005.14165.

Collados Aís, Amparo, and Olga García Becerra. 2015. "Quality." In *Routledge Handbook of Interpreting*, edited by Holly Mikkelson and Renée Jourdenais, 368–83. London: Routledge.

Fantinuoli, Claudio. 2024. "Understanding Machine Interpretation Part 2." Webinar, GALA Global. Accessed February 24, 2024. https://www.gala-global.org/events/events-calendar/understanding-machine-interpretation-part-2.

Fernandes, Patrick, Daniel Deutsch, Mara Finkelstein, Parker Riley, André Martins, Graham Neubig, Ankush Garg, Jonathan Clark, Markus Freitag, and Orhan Firat. 2023. "The Devil Is in the Errors: Leveraging Large Language Models for Fine-Grained Machine Translation Evaluation." In *Proceedings of the Eighth Conference on Machine Translation (WMT)* 1066–83, Tartu, Estonia.

Gieshoff, Anna C., and Michaela Albl-Mikasa. 2022. "Interpreting Accuracy Revisited: A Refined Approach to Interpreting Performance Analysis." *Perspectives* 32(2): 1–19. https://doi.org/10.1080/0907676X.2022.2088296.

Gile, Daniel. 1999. "Variability in the Perception of Fidelity in Simultaneous Interpretation." *Hermes* 22: 51–79. https://doi.org/10.7146/hjlcb.v12i22.25493.

Han, Chao, and Xiaolei Lu. 2021. "Interpreting Quality Assessment Reimagined: The Synergy Between Human and Machine Scoring." *Interpreting and Society* 1(1): 70–90. https://doi.org/10.1177/27523810211033670.

Hendy, Ahmed, Mohamed Abdelrehim, Ahmed Sharaf, Vaibhav Raunak, Mohamed Gabr, Haruto Matsushita, Youngeun Kim, Mohamed Afify, and Hany H. Awadalla. 2023. "How Good Are GPT Models at Machine Translation? A Comprehensive Evaluation." ArXiv. February 18, 2023. https://doi.org/10.48550/arXiv.2302.09210.

Kocmi, Tom, and Christian Federmann. 2023. "Large Language Models Are State-of-the-Art Evaluators of Translation Quality." In *Proceedings of the 24th Annual Conference of the European Association for Machine Translation*, edited by Miikka Nurminen, Jörg Brenner, Maarit Koponen, Sanna Latomaa, Mikhail Mikhailov, Florian Schierl, Tharindu Ranasinghe, Eva Vanmassenhove, Santiago A. Vidal, Naiara Aranberri, Maria Nunziatini, Celia Rico Escartín, Mikel L. Forcada, Maja Popovic, Carolina Scarton, and Helena Moniz, 193–203. Tampere: European Association for Machine Translation. https://aclanthology.org/2023.eamt-1.19.

Komatsu, Tatsuya. 2005. *Tsuyaku no Gijutsu (Interpreting Techniques)*. Tokyo: Kenkyusha.

Kurz, Ingrid. 2001. "Conference Interpreting: Quality in the Ears of the User." *Meta* 46(2): 394–409. https://doi.org/10.7202/003364ar.

Makinae, Mana, Katsuhito Sudoh, Satoshi Nakamura, Kayo Matsushita, and Yuu Yamada. 2023. "Dôji tsûyaku hinshitsu hyôka hôhô kentô no tame no dôji tsûyaku sha to hon'yaku sha no hyôka hikaku bunseki" [A Comparative Analysis of Simultaneous Interpreters' and Translators' Evaluations for the Purpose of Examining Simultaneous Interpretation Quality Assessment Methods]. In *Proceedings of the 29th Annual Meeting of the Association for Natural Language Processing* 29: 1227–81. Okinawa, Japan.

Matsushita, Kayo, and Masaru Yamada. 2022. "Towards the Establishment of a Quality Assessment Framework for Interpreting Performance." Paper presented at Translation in Transition 6, Prague, Czech Republic.

Matsushita, Kayo, and Masaru Yamada. 2023. "How Quality Is Currently Assessed in Human Interpretation and Machine Translation: Towards the Establishment of Quality Assessment for Speech-to-Speech Translation." Paper presented at Symposium on Multi-Modal Automatic Simultaneous Interpretation Research 2023, Nara, Japan.

Matsushita, Kayo, Masaru Yamada, and Hiroyuki Ishizuka. 2020. "An Overview of the Japan National Press Club (JNPC) Interpreting Corpus." *Invitation to Interpreting and Translation Studies* 22: 87–94.

Moral, Rafa, and Janette Mandell. 2023. "A Generative AI Model Outperformed a Neural Machine Translation Engine in One Machine Translation Evaluation: Is This Milestone the Beginning of the End of the Neural Machine Translation Paradigm?" Lionbridge. Last modified May 12, 2023. https://www.lionbridge.com/blog/translation-localization/machine-translation-a-generative-ai-model-outperformed-a-neural-machine-translation-engine/.

Papineni, Kishore, Salim Roukos, Todd Ward, and Wei-Jing Zhu. 2002. "BLEU: A Method for Automatic Evaluation of Machine Translation." In *Proceedings of the 40th Annual Meeting of the Association for Computational Linguistics (ACL)* 311–18. Philadelphia, PA.

Pöchhacker, Franz. 1994. "Quality Assurance in Simultaneous Interpreting." In *Teaching Translation and Interpreting 2*, edited by Cay Dollerup and Annette Lindegaard, 233–42. Amsterdam: John Benjamins.

Pöchhacker, Franz. 2001. "Quality Assessment in Conference and Community Interpreting." *Meta* 46(2): 410–25. https://doi.org/10.7202/003847ar.

Snover, Matthew, Bonnie Dorr, Rich Schwartz, Linnea Micciulla, and John Makhoul. 2006. "A Study of Translation Edit Rate with Targeted Human Annotation." In *Proceedings of the 7th Conference of the Association for Machine Translation in the Americas (AMTA)* 223–31. Cambridge, MA

Someya, Yasumasa. 2017. "A Propositional Representation Theory of Consecutive Notes and Notetaking." In *Consecutive Notetaking and Interpreter Training*, edited by Yasumasa Someya, 145–89. London: Routledge.

Stewart, Craig, Nikolai Vogler, Junjie Hu, Jordan Boyd-Graber, and Graham Neubig. 2018. "Automatic Estimation of Simultaneous Interpreter Performance." In *Proceedings of the 56th Annual Meeting of the Association for Computational Linguistics*, edited by Iryna Gurevych and Yusuke Miyao, 662–66. Melbourne: Association for Computational Linguistics. https://aclanthology.org/P18-2105.

Stojkovski, Bojan. 2023. "Lost and Found in Translation: Is AI Ready to Replace Human Translators?" TheRecursive.com. Last modified April 24, 2023. https://therecursive.com/lost-and-found-in-translation-is-ai-ready-to-replace-human-translators/.

Sutanto, Vincent M., Gian Guido De Giacomo, Toshiaki Nakazawa, and Masaru Yamada. 2024. "ChatGPT as a Translation Engine: A Case Study on Japanese-English." In *Proceedings of the 30th Annual Meeting of the Association for Natural Language Processing* 30: 2096–101. Kobe, Japan.

Wei, Jason, Xuezhi Wang, Dale Schuurmans, Maarten Bosma, Brian Ichter, Fei Xia, Ed Chi, Quoc Le, and Denny Zhou. 2022. "Chain of Thought Prompting Elicits Reasoning in Large Language Models." ArXiv. January 10, 2023. https://doi.org/10.48550/arXiv.2201.11903.

Yamada, Masaru. 2023. "Optimizing Machine Translation through Prompt Engineering: An Investigation into ChatGPT's Customizability." In *Proceedings of Machine Translation Summit XIX* 2: 195–204. Macau SAR, China.

Yamada, Masaru, Kayo Matsushita, and Hiroyuki Ishizuka. 2023. "Utilizing Remote Simultaneous Interpreting Data for Interpreting Quality Assessment: A Corpus-Based Study." In *Corpora in Interpreting Studies: East Asian Perspectives*, edited by Andrew K. F. Cheung, Kanglong Liu, and Ricardo Moratto, 234–50. London: Routledge.

Zhang, Xiaojun. 2016. "Semi-Automatic Simultaneous Interpreting Quality Evaluation." *International Journal on Natural Language Computing* 5(5): 234–50.

Zwischenberger, Cornelia. 2010. "Quality Criteria in Simultaneous Interpreting: An International Vs. a National View." *The Interpreters' Newsletter* 15: 127–42. https://www.openstarts.units.it/handle/10077/4731.

11

COPING WITH CATCH-22

A case study of institutional interpreters' professional ethics in China Mainland

Pan Zhao and Andrew K.F. Cheung

11.1 Introduction

Professional ethics play an essential role in institutional interpreters' work, and failing to properly handle these ethics may result in severe consequences for institutions and institutional leaders. The study, therefore, explores the professional ethics of Chinese interpreters from the perspective of institutions with a case of live simultaneous interpreting for Andrew Parsons in the 2022 Beijing Paralympic Games. Findings demonstrate that the institutional interpreter failed to honor the interpreting ethics of faithfulness but upheld the ethics of institutions, namely, safeguarding the national interest and keeping in line with national policies. The study reveals the tension between interpreting ethics and institutional ethics, particularly the dominance of institutional ethics over interpreting ethics. Moreover, the study showcases that Chinese interpreters' political awareness strongly influences their behaviors in institutional settings. With these findings, the study contributes to the knowledge of Chinese interpreters' professional ethics from the perspectives of institutions and reveals the intricate relations between institutions and interpreters in China.

11.2 Institutional interpreters

Institutional interpreters are those affiliated formally or informally with an institution and bound by institutional norms because they work in the name of, on behalf of, or for the benefit of institutions (Gouadec 2007) such as governments (Guo 2018; Kang 2014), governmental alliances such as the European Parliament (Beaton 2007), courtrooms (Licoppe and Verdier 2013) and even state-owned television (TV) stations (Gu 2019).

DOI: 10.4324/9781003597711-12

When interpreting, institutional interpreters are under three sets of ethics: their interpreting ethics, the ethics of the profession of an institution and their moral values (Jiang 2013). However, the failure to properly handle these ethics during interpreting will not only influence interpreters' behavior or interpreting delivery but also leave repercussions on institutions, such as damaging the image of institutions and institutional leaders or even causing political conflicts (Guo 2015). Therefore, it is urgent to address the institutional interpreters' professional ethics and shed light on the potential ethical dilemma of institutional interpreters. The current study, therefore, explores institutional interpreters' professional ethics with a case of simultaneous interpreting for Andrew Parsons' opening remarks in the 2022 Paralympic Games in Beijing, in which athletes from Russia and Belarus were banned from participating following the Russia-Ukraine military conflicts.

Prior studies on institutional interpreting show that institutions have potential influence over interpreters (Julaiti and Cheung 2024). Within institutional contexts, translators and interpreters are expected to serve as gatekeepers for institutions by delivering institutional values, goals and agendas required by institutions (Kang 2014). For example, Davidson (2000) reveals that the ideology of institutions may affect translation choices, leading to unfaithful translation, and suggests the influence of institutions over translators and interpreters within institutional contexts, but such influence is seldom examined in the context of Chinese institutions.

Within institutional settings, such as the United Nations, the European Parliament, the Hong Kong Legislative Council and the American Institute in Taiwan, interpreters are required to act as gatekeepers by exerting varying degrees of spontaneity (Cheung 2019; Li et al. 2022, 2023; Liu et al. 2023; Ma and Cheung 2020; Wu et al. 2021). In other words, the more spontaneous the source language, the lower the degree of gatekeeping. Therefore, interpreters are known to interpret some heavily scripted speeches by reading out the translated prepared speeches. For example, at the United Nations General Assembly, the Chinese delegation often prepares written translations of Chinese speeches in most if not all of the other official languages for the benefit of interpreters (Wu et al. 2021). However, in the event that these prepared speeches are not provided for interpreters in advance, they will have to rely on oral input, which may compromise gatekeeping (Song and Cheung 2019). At the other end of the spontaneity spectrum, speakers have no prepared documents, and interpreters rely solely on oral input. In these cases, interpreters must be mindful of sensitive expressions and wordings (Li et al. 2022, 2023; Liu et al. 2023; Zhang and Cheung 2022).

The degree of spontaneity is fluid and depends on a number of factors, even within the same intervention. For example, when conducting a clause-by-clause examination of a draft document, a speaker may read the clause under examination from a prepared document but then make spontaneous comments without referring to any documents. In these cases, interpreters must stay alert to the source language while maintaining fluency, as listeners may be critical of non-verbal deviations (Cheung 2013, 2015).

However, such influence of institutions over interpreters, namely gatekeeping and restricted spontaneity, has led to ethical dilemmas for institutional interpreters (Kang 2014). Within the Chinese institutional context, institutional interpreters are particularly affected by the ethics of institutions or, more specifically, the require-ment of political awareness, which the study defines as an all-encompassing term that refers to interpreters' deliberate efforts to minimize political misunderstanding that may result from their target language renditions. However, this finding was based on the self-reports of institutional interpreters and may not be objective and unbiased. Therefore, more updated empirical studies are needed to examine this ethical dilemma of institutional interpreters by contextualizing it in today's Chinese institutions with observation or case studies to draw more objective conclusions.

Therefore, the study explores the professional ethics of Chinese interpreters from the perspective of institutions and elaborates on the ethical dilemmas between institutional interpreters' professional ethics and the ethics of institutions. With the case of simultaneous interpreting of Andrew Parson's opening remarks in the 2022 Beijing Paralympic Games, the study aims to shed light upon the ethical dilemmas faced by institutional interpreters in the unusual one minute of silence during the live broadcast. Besides, the study investigates whether political awareness can still influence institutional interpreters' behaviors in today's China. With the findings, the study expects to enrich the current knowledge of interpreters' professional eth-ics and the power of political awareness in influencing interpreters' behaviors. The study thus offers a peek into the intricate relations between institutions and inter-preters in China.

11.3 Technology-mediated interpreting

Technology-mediated interpreting has garnered significant attention in recent years, with researchers exploring how technology alters or enhances interpreting perfor-mance. Prior studies have examined the integration of computer-assisted interpret-ing tools, the application of machine translation and the growing popularity of remote interpreting platforms (Mellinger and Hanson 2016). For instance, remote interpreting has raised wide concerns about interpreters' cognitive load (Cheung 2024; Julaiti et al. 2024), the potential for communication breakdowns and inter-preting quality (Braun 2015). Moreover, the development of speech recognition software and automatic speech translation tools has been studied in relation to how these technologies complement or challenge human interpreters (Fantinuoli 2018). The advent of more advanced digital tools, such as generative AI, has added to the scholarship that explores the impact of technology on interpreters' performance.

However, extant studies fail to examine another less noticed impact of technol-ogy on interpreting: the ethical issues. For specific occasions that require confi-dentiality, remote interpreting may cause information delivered through unsecured networks and thus run the risk of data breaches (Winteringham 2010). Additionally, interpreters may be constrained by technological conditions that prioritize speed over accuracy, which potentially compromises their ethical obligations to provide faithful interpreting of the source text (ST) (Pym 2016). However, despite there are numerous studies that have addressed general ethical challenges associated with technology use, few have focused on the unique ethical dilemmas, such as faithfulness or neutrality faced by institutional interpreters who work in specific technology-mediated settings.

11.3.1 Live television interpreting: an interpreters' nightmare?

Television interpreting is defined as making foreign language broadcasting content accessible to media users or audiences within the socio-cultural community (Pöchhacker 2016). Assisted by information and communication technologies, traditional conference interpreting has gradually been carried out in macro-contexts, such as televisions. The earliest records of television interpreting in literature date back to the 1950s and 1960s when interpreters were employed for major events in history, such as the opening conference of the International Atomic Energy Agency in 1957, the 1968 US election night and the Apollo space mission to the moon in 1969 (Mayer 1994, 11).

However, television interpreting when being live broadcast is almost unanimously regarded as more stressful than other forms of interpreting (Strolz 1997) because interpreters receive bigger exposure to mass audience in case of failure (Serrano 2011). When interpreting for live-broadcast TV programs, interpreters' delivery is clearly perceived by mass audience and their accidents, if any, may be criticized or analyzed for various purposes. Reasons for television interpreters' accidents abound, notably in verbal or non-verbal aspects. Prior studies have noted linguistic difficulties of TV interpreting, such as the difficulties in interpreting idioms, grammatical inflections, enunciation, pronunciation, comprehension and so on (Darwish 2006). In terms of non-verbal accidents, Ren (2009) defined them as accidents that are not language related but take place during the interpreting process and cause a lack of fluency, such as any emergency during discourse, or backstage conditions.

Be it verbal or non-verbal accidents, live-broadcast television interpreting is frequently referred to the nightmare of television interpreters (Amato and Mack 2011) and raises concerns over interpreters' professional ethics, such as faithfulness. Whether interpreters are able to deliver information accurately on live-broadcast televisions can be perceived by bilingual audience and is further open to scrutinization even after interpreting. Thus, whether interpreters stay faithful to source language on televisions, and the causes of breach of ethics of faithfulness (if any), is of particular interest to scholars and awaits further examination.

11.3.2 Institutional interpreters' mediation

Institutional translators are defined as interpreters affiliated formally or informally with an institution and bound by institutional norms in this study because they work in the name of, on behalf of, or for the benefit of institutions (Gouadec 2007). In this sense, institutional interpreters are preferably characterized by their affiliation with institutions rather than their solid interpreting skills. For example, Ji Chaozhu, a well-known institutional interpreter in the early PRC, was not trained as a professional interpreter but interpreted for several Chinese political leaders at important historical events (Baigorri-Jalón 2017). Therefore, in the context of that time, the definition of institutional interpreting placed greater emphasis on the institutional context of interpreters rather than their interpreting competence. The overt prioritization of political and party loyalty, alongside professional competence, may still linger.

In addition, institutional interpreters are occasionally known to mediate events with dual roles, namely, interpreters and professionals of other fields. As suggested by Heiferman (2014), Madame Chiang not only was the interpreter for her husband, Chiang Kai-Shek, she was also one of the interlocutors at the Cairo Conference. Though a non-professional interpreter, she fulfilled the interpreting task while participating in the conversation. Meanwhile, Ghignoli and Díaz (2015) found that sports commentators for television stations may occasionally double up as interpreters for the audience when live broadcasting the game and can also "serve the purpose" (Ghignoli and Díaz 2015, 206). Previous cases of the dual roles of interpreters and professions of other fields include but are not limited to sports commentators (Ghignoli and Díaz 2015), scholars (Bowen et al. 1995) and even cleaners for various official buildings in hospitals or schools (Pöchhacker and Kadtric 1999). Previous studies found that the dual roles of institutional interpreters and other professionals are not rare in interpreting events (Li and Cheung 2023), and such dual roles have given rise to ethical conflicts between different professions.

As previously explained, interpreters are simultaneously influenced by the code of ethics of interpreting, their moral values and the ethics of institutions (Jiang 2013). Here, the ethics of professions of institutions are defined as the ethical positions of the profession/activity they interpret for (Jiang 2013). For most institutional interpreters, such a dilemma merely refers to the tension between interpreting ethics and the ethics of professions of institutions. For example, Darwish (2006) found that news translators were striking a difficult balance between honoring the core translating ethics of faithfulness and abiding by the biased editorial and institutional policy. Skaaden (2019) also mentioned this phenomenon that the core principles of interpreters' ethics, such as fidelity and impartiality, differentiate their function or role from that of the professionals of institutions they worked for, mostly in public service contexts. However, how interpreters handle the tension between the basic interpreting ethics and the ethics of professionals or institutions, nevertheless, is under-researched in today's academia.

11.3.3 Interpreter's professional ethics

As explained in previous sections, institutional interpreters are sometimes torn between interpreting ethics and their moral values, particularly in community interpreting, public service interpreting (Howes 2023; Pena-Díaz 2018; Baixauli-Olmos 2017), healthcare or legal settings (Remael and Carroll 2015). But the more observed one is the confrontation between the interpreting ethics and the ethics of institutions. Herein, the ethics of institutions are defined as the ethical positions of a profession that interpreters interpret for, where one of the most critical determinants in this ethical sphere is the institutional setup and the purpose of the institution's activities (Jiang 2013). In legal settings, it is the professional ethics of lawyers or law authorities; for mass media institutions, it is the ethics of mass media professionals. In one typical study by Englund Dimitrova (2019), the interpreter was also the journalist, and the ethics of journalism dominated the ethics of interpreters and influenced the interpreter's delivery accuracy. This study demonstrates that the ethics of institutions may clash with the interpreters' professional ethics, which again explains why institutional interpreters face an ethical dilemma between these two sets of ethics.

Likewise, Chinese researchers have also noted the ethical dilemma between interpreters' professional ethics and the ethics of institutions. Ren (2020) found early institutional interpreters in China were subject to powerful fields such as politics and diplomacy and were expected to honor fidelity, integrity and political awareness. During this stage, interpreters were expected to be neither transparent nor neutral. Ren's study (2020) also points out the fact that early institutional interpreters were torn between common interpreting ethics and the requirements of political awareness required by institutions. However, in-depth scholarly studies regarding specific interpreters' professional ethics remain scant (Ren 2020). More research is, therefore, needed to address specific interpreters' professional ethics and such dilemmas with more fine-grained studies within more specific contexts.

11.3.4   *Political awareness of Chinese institutional interpreters*

As suggested in the previous section, Chinese institutional interpreters are influenced by the ethics of institutions, one of which is the requirement of political awareness. Here, political awareness is defined as a form of socio-institutional cognition that results from understanding socio-institutional requirements (Guo 2018), institutional policy and even national policy.

Studies have shown that Chinese institutional interpreters' strategies can be manipulated by their political awareness. Guo (2015) discussed three types of interpreting strategies manipulated by institutional interpreters' political awareness: the addition of experiential meaning to express a political standpoint, the omission of experiential meaning to eliminate potential adverse political effects and the correction of inaccurate experiential meaning to avoid political misunderstanding. By citing examples from interpreting previous Premier Zhu's debut speech, Guo (2015) explained how political awareness can influence and manipulate interpreters' behaviors. However, the study discussed a case of previous Premier Zhu's debut speech which was made 24 years ago. Therefore, it is unclear whether political awareness can still influence interpreters' behaviors and their delivery in today's Chinese institutional contexts. Future scholars, therefore, are advised to study how political awareness influences interpreters' behaviors in more updated institutional contexts in today's China.

It can be seen from previous reviews that extant studies have not adequately addressed the intricate relations between interpreters and institutions and the ethical dilemmas faced by institutional interpreters. Within the Chinese institutional context, such a dilemma refers to the confrontation between the ethics of institutions and interpreters' professional ethics, mostly faithfulness. Moreover, it is unknown whether political awareness still influences today's institutional interpreters' behavior. Therefore, this study seeks to examine Chinese institutional interpreters' professional ethics and address this ethical dilemma with the case of the 2022 Beijing Paralympic Opening Ceremony. In particular, the study explores whether the interpreters' professional ethic of faithfulness was upheld during interpreting and if institutional ethics, such as political awareness, influenced interpreters' behaviors. Therefore, two research questions are thus raised based on the previous review:

Rq1: To what extent did the institutional interpreter for the 2022 Beijing Paralympics honor the professional ethics of faithfulness specified by the Translators Association of China? How did institutional ethics influence his behaviors?

Rq2: To what extent was the institutional interpreter for the Paralympics subject to political awareness when interpreting politics-related messages? If so, how did political awareness influence his behaviors?

11.4 Materials and methods

This is a case study of the simultaneous interpreting of Andrew Parsons' opening remarks in the 2022 Beijing Paralympic Games. The game started at 20:00 on March 4, 2022, and was broadcast live in China mainland and the rest of the world. The opening statement by Andrew Parsons, the current president of the International Paralympic Committee (IPC), was interpreted simultaneously into Mandarin Chinese to the mainland audience. His speech lasted 5 minutes and 52 seconds and was interpreted by a male interpreter simultaneously from English to Mandarin Chinese and a sign language interpreter. However, the interpreter stopped interpreting at 00:49 of the speech and resumed at 01:55. Besides, the interpreter also interpreted one sentence from English to Mandarin Chinese one minute earlier than when the speaker should utter it. The researcher transcribed the dis-interpreted part of the speech into texts for analysis.

The speech made by Andrew Parsons lasted 5 minutes and 52 seconds, and his speech was interpreted by a male interpreter simultaneously from English to Chinese on TV and a female sign language interpreter who appeared on the right-down corner of the screen. At 00:57, for unknown reasons, the interpreter stopped interpreting and resumed interpreting at 01:55. The omitted ST is listed below:

> At the IPC, we aspire to a better and more inclusive world, free from discrimination, free from hate, free from ignorance, and free from conflicts. Here in Beijing, Paralympic athletes from 46 different nations will compete with each other, not against each other. Through sports, they will showcase the best of humanity and highlight the values that should underpin a peaceful and inclusive world. Paralympians know that opponent does not have to be an enemy and that united, we can achieve more, much more.

11.5 Results

Compared with other parts of the speech, the part of ST is believed to deliver a peace-appealing message. The words such as "inclusive world", "free from discrimination, free from hate, free from ignorance and free from conflicts", "peaceful and inclusive world" and "united" are considered to be evidence for the assumption. In this short paragraph, the word "inclusive" appears two times, and the word "free from" appears four times, thus considered the keywords of the paragraph.

Without any explanation for this one-minute silence, the event leaves researchers with no first-hand answer but making speculations about the interpreter's performance. The peace-appealing message in the speech may primarily be attributed to the value of the Paralympics. Therefore, it is likely that Andrew Parsons, the president of IPC, was trying to promote the IPC's value of social inclusion by repeating the keywords of inclusion in his speech. In addition, the message seems to be a rhetorical response to the then-military conflicts between Russia and Ukraine in early 2022. The military conflict between Russia and Ukraine started on February 24th, 2022, with Russia's missile strikes over Ukraine. Deemed an "invader" in many presses, such as BBC (British Broadcast Corporations [BBC] 2024), Russia has been under criticism and sanctions since its military conflicts with Ukraine began. On the official website of IPC, an opening statement was made on March 2nd, 2022, stating that the Paralympic delegations from Russia and Belarus will participate as neutrals and will not be included in the medal table. The statement of IPC is considered as its stance over the Russia-Ukraine military conflicts or its sanction against the Russian government. Although Andrew Parsons did not express his opinions about the conflict or mention anything about it, he likely appealed for peace as a response to the then-military conflict between Russia and Ukraine. However, until today, there have been no official explanations given by the interpreter or the organizing committee, which makes it impossible for the researcher to track the original cause of the omission in the Chinese interpretation.

11.6 Discussion

The study addresses professional ethics and the ethical dilemmas of institutional interpreters in China mainland. In particular, the study explores whether the interpreter's behavior honored the Translators Association of China's (TAC) professional ethics of faithfulness and whether the ethics of institutions or political awareness influenced the interpreter. By reflecting on the one-minute silence in the interpreting process, the researcher aims to gain a deep understanding of the professional ethics of interpreters in China mainland.

In the official document Code of Professional Ethics for Translators and Interpreters (Translators Association of China [TAC] 2019), article 5.4.1.1 stipulates that interpreters should present faithful translation or interpretation of the ST except for court, psychological therapy and similar settings. Meanwhile, Moody (2011) pointed out that faithful interpretation means "we should not interject our opinions, nor should we add or subtract anything from the message" (Moody 2011, 1). Following these two theories, it is found that the interpreter failed to present a faithful interpretation or fully uphold the professional ethics of faithfulness.

Furthermore, this television and institutional interpreter is subject to not only professional interpreting ethics but also professional ethics of television network. As explained in the previous paragraph, interpreters are influenced by three sets of ethics during interpreting: personal moral values, professional ethics of interpreting and ethics of independent professions of institutions (Jiang 2013). In this case, the interpreter worked in the government-affiliated events on state-owned televisions and for the benefit of the institutions and is thus categorized as an institutional and television interpreter following the definition of Gouadec (2007). In this case, this institutional interpreter played the dual roles of both an interpreter and a mass media professional (TV anchor). Therefore, it is debatable whether the interpreter stopped interpreting not to honor interpreters' professional ethics of faithfulness but to be guided by the ethics of mass media professionals.

Professional ethics for institutions in radio and television industries are stipulated in Proposal for the Construction of Radio and Television Professional Ethics (China Federation of Radio and Television Associations 2021). Article 2.1 of this document states that the workers in this industry should uphold the leadership of the Communist Party of China and national interest and must not publish or disseminate any information that damages the party's image or the country. This brings the story to a larger context. One of China's foreign policies is that China will not intervene in the internal affairs of other countries, nor will the Chinese government recklessly comment on the internal affairs of other countries. In a government-affiliated event broadcast on state-owned television, one of the primary professional ethics of mass media professionals is to safeguard the national interest and to keep in line with government policies. Therefore, although the interpreter stopped interpreting and failed to uphold interpreters' ethics of faithfulness, the interpreter honored the professional ethics of mass media by keeping silent and stopping delivering peace-appealing messages in public, which may otherwise be interpreted by foreign media as taking sides in Russia-Ukraine conflicts and breaching China's foreign policies. In other words, the institutional interpreter chose to honor the institutional ethics of mass media and safeguarded China's national interest to such an extent that he could not simultaneously uphold the interpreter's ethics of faithfulness.

Interestingly, if the previous deduction makes sense, the case did highlight the conflicts between two sets of professional ethics for institutional interpreters in China mainland, namely, the ethics of interpreting and the ethics of mass media institutions. This lends support to previous similar studies of Drugan (2017), who discussed the case when interpreters had to perform by both the ethics of social workers and the ethics of interpreters, while the ethics of interpreters prevented the fulfillment of the ethics of social workers. Another similar study conducted by Englund Dimitrova (2019) discussed such a phenomenon of clashing ethics for a talk show host, who, at the same time, was also the interpreter for the show. Two sets of ethics influenced his behaviors simultaneously and the host's role dominated over the interpreter's role. Consistent with their finding, the study reaches a similar finding that the ethics of mass media dominated over the ethics of interpreters in this one minute of silence.

Meanwhile, such a conflict between two sets of professional ethics also mirrors the influence of institutions over interpreters. From the perspective of institutions, interpreters are not only bound to interpreting ethics but are also expected to honor institutional ethics. In this sense, institutional interpreters are heavily under the influence of institutions and thus take complicity with institutions during the meaning-making process (Wallmach 2014). Institutional interpreters can no longer be a neutral conduit of information exchange but also be constrained by institutions by delivering the wanted voice and being restricted by institutional ethics. Such a finding is consistent with the previous study of Schäffner et al. (2014), who claimed that institutional translators and interpreters speak for and reflect the goals and voice of the institutions they work for.

In this case, the institutional interpreter's act of stopping interpreting for one minute can be seen as the omission of information, which may be explained from the perspective of in institutional political awareness. Guo (2015) once posited that omission can be utilized in political-institutional contexts to eliminate potential adverse political effects and avoid the experiential meaning that may affect the leader's and institution's image. In this sense, the interpreter for Premier Zhu acted not only as the conduit of information but also as the guardian of the Chinese state leader's and government's image. Consistent with the finding of Guo (2015), this study considers the omission during interpreting as a strategy to avoid political misunderstanding that would otherwise be misinterpreted as breaching China's foreign policy by delivering the Chinese government's voice or stance on the controversial Russia-Ukraine conflict. In other words, the act of stopping interpreting is believed to be guided by the institutional interpreter's awareness of politics, namely, his knowledge of Chinese foreign policy and the military conflicts between Russia and Ukraine.

Interestingly, if the previous deduction makes sense, the finding does reveal such political awareness as an unspoken rule governing institutional interpreters' behaviors, which is nowhere seen in the interpreting ethics guidelines issued by TAC. This unspoken rule of today's institutional interpreters echoes the earlier self-reports of Chinese diplomatic interpreters for the Foreign Ministry, who claimed that the primary qualification of an interpreter is his or her "command of politics" (Ren 2020, 281). Today's Chinese institutional interpreters are still greatly subject to such political awareness despite the tremendous change in the socio-political contexts of China in the past 70 years. Though such political awareness is never written in the official documents that address interpreting ethics, it is an unspoken rule that influences institutional interpreters in today's China.

11.7 Conclusion

This case study investigates the professional ethics of Chinese interpreters from the perspective of institutions. Focusing on the one-minute silence, the study suggests that the interpreter did not honor the interpreter's ethics of faithfulness to such an extent that the interpreter stopped interpreting. However, the interpreter simultaneously upheld the institutional ethics of mass media in China mainland, or more specifically, political awareness.

The study highlights institutions' strong influence over interpreters in an institutional setting. As demonstrated in the study, when working in institutional settings, interpreters are no longer independent entities separated from institutions but are restricted by institutions. Within Chinese institutional settings, one example of such restriction is political awareness. Political awareness still exerts great influence on today's institutional interpreters in China. The study, therefore, introduces institutions as a novel perspective to interpreters' ethics of Chinese interpreters and expounds on the relations between institutions and interpreters.

Moreover, the study sheds light on the ethical dilemmas faced by institutional interpreters in China. Interpreters' interpreting ethics and ethics of institutions co-exist when exerting influence on institutional interpreters, and institutional ethics play a more dominant role than the interpreters' interpreting ethics. This finding lends support to previous claims of Englund Dimitrova (2019), who posited that the role of institutions professionals may sometimes overshadow the role of interpreters. However, the study fails to explain the cause of the dilemma and thus creates a gap that future studies are expected to fill.

The study also offers an alternative approach to investigating interpreters' ethics by initiating an open-ended discussion about ethics rather than scrutinizing ethics guidelines. Different from previous studies that took interpreters' ethics as stiff doctrines, the study approaches interpreters' ethics as principles that are open to examination. Consistent with the finding of Pöchhacker (2015), the shift from a more universalist statement to a more relativistic attitude demonstrated in this study pushes future studies on interpreters' ethics into more open-ended discussions. The statement means future studies on this topic can be conducted with more empirical studies in various contexts, which can help expand the boundary of studies beyond which interpreters' ethics can be explained.

However, until today, there has been no official explanation for the silence during the live broadcasting of interpreting. An accurate understanding of what happened during the one minute can never be sought but sealed in history. Since history is invariably penned by people and inevitably subjected to prejudice and bias, a post-interpreting understanding can hardly draw any firm conclusion. This explains why the conclusion in the study is more deductive than conclusive.

Meanwhile, it remains doubtful whether the findings from the single case can be generalized and applied to other institutional contexts in China. The difficulty in accessing the record of similar events makes it almost impossible to conduct clustered studies and draw a more solid conclusion. More empirical studies, therefore, are needed to shed light on institutional interpreters' professional ethics, particularly their ethical dilemmas, and examine the role of political awareness in influencing institutional interpreters' behaviors while interpreting.

References

Amato, Amalia Agata Maria, and Gabriele Mack. 2011. "Interpreting the Oscar Night on Italian TV: An Interpreters' Nightmare?" *The Interpreters' Newsletter* 16: 37–60.

Baigorri-Jalón, Jesús. 2017. *Autobiography or History? Ji Chaozhu, an Interpreter in Mao's China.* Berlin: Frank & Timme.

Baixauli-Olmos, Lluís. 2017. "Ethics Codes as Tools for Change in Public Service Interpreting: Symbolic, Social and Cultural Dimensions." *The Journal of Specialised Translation* 28(8): 250–72.

Beaton, Morven. 2007. Intertextuality and Ideology in Interpreter-Mediated Communication: The Case of the European Parliament." PhD dissertation, Heriot-Watt University, Edinburgh.

Bowen, Margareta, David Bowen, Francine Kaufmann, and Ingrid Kurz. 1995. "Interpreters and the Making of History." In *Translators through History*, edited by Jean Delisle and Judith Woodsworth, 245–73. Amsterdam: John Benjamins.

Braun, Sabine. 2015. "Remote Interpreting." In *The Routledge Handbook of Interpreting*, edited by Holly Mikkelson and Renée Jourdenais, 352–67. New York: Routledge.

British Broadcast Corporations (BBC). 2024. "Ukraine in Maps: Tracking the War with Russia." November 19, 2024. https://www.bbc.com/news/world-europe-60506682.

Cheung, Andrew K. F. 2013. "Non-Native Accents and Simultaneous Interpreting Quality Perceptions." *Interpreting* 15(1): 25–47. https://doi.org/10.1075/intp.15.1.02che.

Cheung, Andrew K. F. 2015. "Scapegoating the Interpreter for Listeners' Dissatisfaction with Their Level of Understanding: An Experimental Study." *Interpreting* 17(1): 46–63. https://doi.org/10.1075/intp.17.1.03che.

Cheung, Andrew K. F. 2019. "The Hidden Curriculum Revealed in Study Trip Reflective Essays." In *American Translators Association Scholarly Monograph Series*, edited by David B. Sawyer, Frank Austermühl, and Vanessa Enríquez Raído, 393–408. Amsterdam: John Benjamins Publishing Company.

Cheung, Andrew K. F. 2024. "Cognitive Load in Remote Simultaneous Interpreting: Place Name Translation in Two Mandarin Variants." *Humanities and Social Sciences Communications* 11(1): 1–8. https://doi.org/10.1057/s41599-024-03767-y.

China Federation of Radio and Television Associations. 2021. "Proposal for the Construction of Radio and Television Professional Ethics." March 30, 2021.

Darwish, Ali. 2006. "Translating the News Reframing Constructed Realities." *Translation Watch Quarterly* 2(1): 52–77.

Davidson, Brad. 2000. "The Interpreter as Institutional Gatekeeper: The Social-Linguistic Role of Interpreters in Spanish-English Medical Discourse." *Journal of Sociolinguistics* 4(3): 379–405. https://doi.org/10.1111/1467-9481.00121.

Drugan, Joanna. 2017. "Ethics and Social Responsibility in Practice: Interpreters and Translators Engaging with and Beyond the Professions." *Translator* 23(2): 126–42. https://doi.org/10.1080/13556509.2017.1281204.

Englund Dimitrova, Birgitta. 2019. "Changing Footings on 'Jacob's Ladder': Dealing with Sensitive Issues in Dual-Role Mediation on a Swedish TV Show." *Perspectives, Studies in Translatology* 27(5): 718–31. https://doi.org/10.1080/0907676X.2018.1524501.

Fantinuoli, Claudio. 2018. "Interpreting and Technology: The Upcoming Technological Turn." In *Interpreting and Technology*, 1–12. Doi: 10.5281/zenodo.1493289

Ghignoli, Alessandro, and María Gracia Torres Díaz. 2015. "Interpreting Performed by Professionals of Other Fields: The Case of Sports Commentators." In *Non-Professional Interpreting and Translation in the Media*, edited by Rachele Antonini and Chiara Bucaria, 193–208. Frankfurt am Main: Peter Lang.

Gouadec, Daniel. 2007. *Translation as a Profession*. Amsterdam: John Benjamins.

Gu, Chonglong. 2019. "(Re)Manufacturing Consent in English: A Corpus-Based Critical Discourse Analysis of Government Interpreters' Mediation of China's Discourse on PEOPLE at Televised Political Press Conferences." *Target: International Journal of Translation Studies* 31(3): 465–99. https://doi.org/10.1075/target.18023.gu.

Guo, Yijun. 2015. "The Interpreter's Political Awareness as a Non-Cognitive Constraint in Political Interviews: A Perspective of Experiential Meaning." *Babel* 61(4): 573–88. https://doi.org/10.1075/babel.61.4.07yij.

Guo, Yijun. 2018. "Effects of the Interpreter's Political Awareness on Pronoun Shifts in Political Interviews: A Perspective of Interpersonal Meaning." *Babel* 64(4):528–47. https://doi.org/10.1075/babel.00053.guo.

Heiferman, Ronald Ian. 2014. *The Cairo Conference of 1943: Roosevelt, Churchill, Chiang Kai-shek and Madame Chiang*. Jefferson: McFarland & Company.

Howes, Loene M. 2023. "Ethical Dilemmas in Community Interpreting: Interpreters' Experiences and Guidance from the Code of Ethics." *The Interpreter and Translator Trainer* 17(2): 264–81. https://doi.org/10.1080/1750399X.2022.2141003.

Jiang, Hong. 2013. "The Ethical Positioning of the Interpreter." *Babel* 59(2): 209–23. https://doi.org/10.1075/babel.59.2.05jia.

Julaiti, Kaifusai, and Andrew K. F. Cheung. 2024. "Two Tales of a City: Simultaneous and Consecutive Interpreting in Hong Kong." In *The Routledge Handbook of Chinese Interpreting*, edited by Riccardo Moratto and Cheng Zhan, 368–78. London: Routledge.

Julaiti, Kaifusai, Nina Delia Y. Y. Cheung, Andrew K. F. Cheung, and Jessy Yujie Huang. 2024. "Number Training in Simultaneous Interpreting: A Corpus-Assisted Longitudinal Study." *Interpreting and Society*. https://doi.org/10.1177/27523810241285542.

Kang, Ji-Hae. 2014. "Institutions Translated: Discourse, Identity and Power in Institutional Mediation." *Perspectives, Studies in Translatology* 22(4): 469–78. https://doi.org/10.1080/0907676X.2014.948892.

Li, Danni and Andrew K. F. Cheung. 2023. "Simultaneous Interpreting of Online Medical Conferences: A Corpus-Based Study." In *Corpora in Interpreting Studies*, edited by Andrew K. F. Cheung, Kanglong Liu, and Riccardo Moratto, 115–32. London: Routledge.

Li, Ruitian, Andrew K. F. Cheung, and Kanglong Liu. 2022. "A Corpus-Based Investigation of Extra-Textual, Connective, and Emphasizing Additions in English-Chinese Conference Interpreting." *Frontiers in Psychology* 13: 844735. https://doi.org/10.3389/fpsyg.2022.847735.

Li, Ruitian, Kanglong Liu, and Andrew K. F. Cheung. 2023. "Interpreter Visibility in Press Conferences: A Multimodal Conversation Analysis of Speaker–Interpreter Interactions." *Humanities and Social Sciences Communications* 10(1): 454–12. https://doi.org/10.1057/s41599-023-01974-7.

Licoppe, Christian, and Maud Verdier. 2013. "Interpreting, Video Communication and the Sequential Reshaping of Institutional Talk in the Bilingual and Distributed Courtroom." *The International Journal of Speech, Language & the Law* 20(2): 247–75. https://doi.org/10.1558/ijsll.v20i2.247.

Liu, Yi, Andrew K. F. Cheung, and Kanglong Liu. 2023. "Syntactic Complexity of Interpreted, L2 and L1 Speech: A Constrained Language Perspective." *Lingua* 286: 103509. https://doi.org/10.1016/j.lingua.2023.103509.

Ma, Xingcheng, and Andrew K. F. Cheung. 2020. "Language Interference in English-Chinese Simultaneous Interpreting with and Without Text." *Babel* 66(3): 434–56.

Mayer, Horst F. 1994. "Live Interpreting for Television and Radio." *The Jerome Quarterly* 9(2): 11.

Mellinger, Christopher, and Thomas Hanson. 2016. *Quantitative Research Methods in Translation and Interpreting Studies*. London: Routledge. https://doi.org/10.4324/9781315647845.

Moody, Bill. 2011. "What is a Faithful Interpretation?" *Journal of Interpretation* 21(1).

Pena-Díaz, Carmen. 2018. "Ethics in Theory and Practice in Spanish Healthcare Community Interpreting." *Monografías de traducción e interpretación* 10.

Pöchhacker, Franz. 2015. *Routledge Encyclopedia of Interpreting Studies*. London: Routledge.

Pöchhacker, Franz. 2016. *Introducing Interpreting Studies*. London: Routledge.

Pöchhacker, Franz, and Mira Kadric. 1999. "The Hospital Cleaner as Healthcare Interpreter: A Case Study." *Translator* 5(2): 161–78. https://doi.org/10.1080/13556509.1999.10799039.

Pym, Anthony. 2016. *Translation Solutions for Many Languages: Histories of a Flawed Dream*. London: Bloomsbury Publishing.

Remael, Aline, and Mary Carroll. 2015. "Community Interpreting: Mapping the Present for the Future." *The International Journal of Translation and Interpreting Research* 7(3): 1–9.

Ren, Wen. 2009. *Jiaoti chuanyi* 交替传译 [Consecutive Interpretation]. Beijing: Foreign Language Teaching and Research Press.

Ren, Wen. 2020. "The Evolution of Interpreters' Perception and Application of (Codes of) Ethics in China Since 1949: A Sociological and Historical Perspective." *Translator* 26(3): 274–96. https://doi.org/10.1080/13556509.2020.1832019.

Schäffner, Christina, Luciana Sabina Tcaciuc, and Wine Tesseur. 2014. "Translation Practices in Political Institutions: A Comparison of National, Supranational, and Non-Governmental Organisations." *Perspectives, Studies in Translatology* 22(4): 493–510. https://doi.org/10.1080/0907676X.2014.948890.

Serrano, Óscar Jiménez. 2011. "Backstage Conditions and Interpreter's Performance in Live Television Interpreting: Quality, Visibility and Exposure." *The Interpreters' Newsletter* 16: 115–36.

Skaaden, Hanne. 2019. "Invisible or Invincible? Professional Integrity, Ethics, and Voice in Public Service Interpreting." *Perspectives, Studies in Translatology* 27(5): 704–17. https://doi.org/10.1080/0907676X.2018.1536725.

Song, Shuxian, and Andrew K. F. Cheung. 2019. "Disfluency in Relay and Non-Relay Simultaneous Interpreting: An Initial Exploration." *FORUM* 17(1): 1–19. https://doi.org/10.1075/forum.18016.che.

Strolz, Birgit. 1997. "Quality of Media Interpreting: A Case Study." In *Conference Interpreting: Current Trends in Research*, edited by Yves Gambier, Daniel Gile, and Christopher Taylor, 194–97. Amsterdam: John Benjamins.

Translators Association of China (TAC). 2019. "Code of Professional Ethics for Translators and Interpreters." https://www.tac-online.org.cn/node_1015757.html.

Wallmach, Kim. 2014. "Recognising the 'Little Perpetrator' in Each of Us: Complicity, Responsibility and Translation/Interpreting in Institutional Contexts in Multilingual South Africa." *Perspectives, Studies in Translatology* 22(4): 566–80. https://doi.org/10.1080/0907676X.2014.948893.

Winteringham, Sarah Tripepi. 2010. "The Usefulness of ICTs in Interpreting Practice." *The Interpreters' Newsletter* (15): 87–99.

Wu, Baimei, Andrew K. F. Cheung, and Jie Xing. 2021. "Learning Chinese Political Formulaic Phraseology from a Self-Built Bilingual United Nations Security Council Corpus: A Pilot Study." *Babel* 67(4): 500–21. https://doi.org/10.1075/babel.00233.wu.

Zhang, Yifan, and Andrew K. F. Cheung. 2022. "A Corpus-Based Study of Modal Verbs in Chinese–English Governmental Press Conference Interpreting." *Frontiers in Psychology* 13: 1065077. https://doi.org/10.3389/fpsyg.2022.1065077.

12

SELF-ASSESSMENT FOR SIMULTANEOUS INTERPRETING

The influence of speech-to-text technology and expert demonstration

Zi-ying Lee

12.1 Introduction: technology and interpreter training

For decades, professional conference interpreters have been using technology to support their work. During the preparation stage, technology or more specifically, computer-assisted interpreting tools as referred to by Fantinuoli (2017, 2022) are used to support the interpreters at both linguistic and extralinguistic levels as they gather background knowledge, find texts, papers, audio and visual records of the speakers and manage terminology. During the conference, in addition to using their laptops to read the slides, papers and list of guests and search for unfamiliar terms, interpreters nowadays can also choose to use speech-to-text technology (STT) or automatic speech recognition as an aid to support their interpreting (Corpas Pastor 2018; Defrancq and Fantinuoli 2021; Fantinuoli 2017, 2022; Mellinger 2019; Ortiz and Cavallo 2018; Pisani and Fantinuoli 2021; Zhao 2022).

Seeing the benefits of technology, interpreter trainers have also taught student interpreters to harness the potential of technology. The idea of computer-assisted interpreter training (CAIT) appeared as early as in the 1990s when interpreter trainers began to experiment using multimedia as well as information and communication technology (ICT) to support their training of interpreters (Carabelli 1999; Cervato and de Ferra 1995; Deysel and Lesch 2018). Over the years, trainers mainly took three approaches to CAIT: (1) creation of speech databases or speech repositories; (2) designing computer programs for interpreter trainers to create exercises; and (3) virtual learning environments (VLEs) for interpreter training (Sandrelli and Jerez 2007).

The first approach later combines speech banks with corpus and creates large speech repositories like European Speech Repository (Bendazzoli and Sandrelli 2005; Monti et al. 2005; Sandrelli and Jerez 2007). The second resulted in programs

DOI: 10.4324/9781003597711-13

like Interprit (Cervato and de Ferra 1995; Merlini 1996) and Blackbox (Sandrelli 2005). The third approach involves course management system and blended learning with materials made available for students to carry out self-study, peer review and self-assessment (Chang and Hao 2008; Class and Moser-Mercer 2006, 2013; Kajzer-Wietrzny and Tymczynska 2014).

As stated by Fantinuoli (2022), CAIT enables trainers to utilize ICT to enhance the training of interpreters as interpreter training moves from a teacher-centered toward a learner-centered approach. One of the areas that CAIT may be useful is to support student interpreters' self-study and self-assessment (Frittella 2021).

12.2 Deliberate practice and self-assessment

Interpreter training has always stressed the importance of deliberate practice and self-study (Ericsson et al. 1993; Motta 2011; Tiselius 2013, 2018). According to Ericsson and Harwell (2019), deliberate practice should (1) involve a trainee and a teacher who can *assess* what areas the trainee needs to work on and recommend practice method; (2) the *goals* of practice should be made clear between the teacher and the trainee; (3) the practice activity should allow the trainee to get *feedback* to tell the differences between current performance and desired goals; and (4) repeated and revised attempts of the trainee (Ericsson and Harwell 2019, italic added). However, when student interpreters engage in self-study, it is not easy or even possible to fulfill all the above criteria, particularly when it comes to self-assessment. Nevertheless, scholars generally hold the view that deliberate practice is essential for learners to master a certain skill (see, for instance, Ericsson and Harwell 2019; Motta 2011; Tiselius 2013, 2018).

Having said that, interpreter trainers also understand the risks of having students carry out self-study without proper reflection and self-assessment (Deysel and Lesch 2018). As stated by Sandrelli (2005), "if unsupervised practice sessions are to be useful, students need to be able to assess their own performance and identify their weaknesses. Indeed, the development of self-assessment skills is an essential component of interpreter training" (195).

To enable student interpreters to review their own performance during self-study and identify areas that need to be fixed, interpreter trainers and scholars have developed self-assessment sheets and reflection logbooks (Deysel and Lesch 2018; Fowler 2007; Han and Fan 2020; Hartley et al. 2003; Lee 2005; Lee, 2016, 2022; Li 2018; Liu 2021; Tang and Ju 2023). Another common approach is to have students transcribe their own interpretation (Yang et al. 2023) and search for omissions or meaning errors in their interpretation for reflection and revision as they repeatedly use the same materials for practice. Nevertheless, without the presence of the trainer, it is not easy for student interpreters to identify or diagnose their problems and self-assessment for interpreting can be time-consuming (Lee 2022). Studies have also shown that even with self-assessment sheet (Han 2018; Han and Fan 2020; Lee 2017), the effect of self-assessment varies from person to person.

At the same time, student interpreters may worry that their assessment criteria may be different from their teachers or question the quality of peer assessment or self-assessment (Lee 2018).

12.3 The project

12.3.1 The context

Based on the author's personal experience of teaching simultaneous interpreting, it is quite common for student interpreters to encounter various challenges when they first start to learn simultaneous interpreting. These challenges include difficulty of multi-tasking; unable to prioritize messages; incomplete sentences; lack of background knowledge; unable to cope with speeches with fast speed or strong accents and so on. As the course progresses, students may become more and more frustrated.

Courses of simultaneous interpreting usually require the use of specially designed classrooms equipped with interpreting booths or equipment, so the number of students who can take the course is limited. Nevertheless, as pointed out by Lee (2018), it can be quite time-consuming for trainers to give detailed feedback on individual student's performance, particularly for undergraduate program when the number of students may exceed 20. The situation is better for graduate program with about ten students, but graduate students will face more challenges as the teaching materials tend to be more difficult compared with those used in undergraduate program.

To compensate for the limited time allowed for individual feedback, the author would ask student interpreters to transcribe their interpretation of a class practice or the midterm exam for self-assessment and reflection and students are expected to critique on their peers' performance. However, over the years, students often complain that transcribing is extremely time-consuming as they need to use nearly one hour to transcribe their interpretation that lasts only three to five minutes, plus self-assessment and reflection.

12.3.2 Research questions

As STT is getting more mature and reliable, students nowadays can use STT software or mobile apps to generate transcripts of their interpretation. The first goal of this study, thus, is to explore if STT can effectively support students when they engage in self-assessment of simultaneous interpreting.

Second, even though trainers will introduce various strategies of simultaneous interpreting in class, it is not easy for students to know when to apply these strategies. In my class, students usually have a better understanding about the ways to apply these strategies as well as the suitable timing after they hear a demonstration. Studies have also shown that expert interpreters and student interpreters use strategies differently (Arumí Ribas 2012; Díaz-Galaz et al. 2015; Li 2019). Hence, the second goal of this study is to provide students with audio files and transcripts of demonstration interpretation done by professional conference interpreters and see if such demonstration can enable students to grasp the essence of different interpreting strategies and how to apply them.

In short, this study aims to explore two research questions:

Rq1: To what extent can STT support student interpreters when they engage in self-assessment of simultaneous interpreting?

Rq2: When STT is used in combination with expert demonstration resources (including audio files and transcripts), to what extent and in which aspects can they help boost student interpreters' confidence when they carry out self-assessment?

12.3.3 The course

The project was carried out in 2021 in a Taiwanese university that offered interpreting courses to both undergraduate and graduate students. Two courses of simultaneous interpreting were used for this study. The basic course design is very similar but one course was offered as an elective course for senior college students and the other was offered as an elective course to second-year graduate students. Each course lasted 18 weeks and information regarding this study was distributed at the start of the first week to get students' consents. In the end, 25 senior college students and seven graduate students participated in this study.

In the first three weeks, fundamental skills related to simultaneous interpreting, such as sight translation and shadowing, were introduced. Then students were guided to learn and practice simultaneous interpreting with general texts from week 4 to week 6. In the meantime, assessment criteria as well as differences between assessing simultaneous interpreting and consecutive interpreting were introduced. As pointed by Lee (2005), for students to engage in self-assessment, it is important for the instructor to ensure that students clearly understand the criteria.

From the fifth week to the eighth week, students were instructed to engage in self-practice after class and an STT platform, Yating ASR, was provided for them to transcribe their recording. Students were required to submit at least one audio file and the corresponding transcript to ensure that students know how to use the software and how to compare their interpretation against the source text provided by the author.

At the beginning of the semester, the author selected four speeches to be used from week 10 to week 17 for undergraduate students and four speeches for graduate students and four professional conference interpreters were invited to provide the demonstration interpreting audio. The STT platform was used to generate the transcript with the timecode of both the source speech and the expert interpretation. The interpreters were asked to prepare for the interpreting assignment as if it was a real assignment and provide their interpretation. The author also told them that mistakes were allowed as mistakes do happen in real situations.

From the tenth week, information related to the selected source speech, including the speaker's bio, the PowerPoint slides (if available) and so on, were provided to the students for them to search for relevant information and prepare a glossary list. Then, each speech was divided into two. The first half was used in class. Recordings of students' interpreting in class were used to discuss issues related to accuracy, completeness, fluency, breathing and expressions. After the class, the audio file was made available on the VLE platform for students to practice at home and submit their recordings (preferably after repeated practice) before the designated deadline. Expert demonstration resources (i.e. the audio and the transcript with timecode) were made available after all students submitted their assignments. Students would then use their own recording and transcript generated by Yating to compare with the experts' version and engage in self-assessment on accuracy, completeness, fluency and strategies.

In the next class, with the author as the facilitator, students were guided to discuss challenges, problems identified and what they have seen from the expert demonstration. The remaining second half of the speech was then used in class. For the second half, audio file and expert demonstration were provided on the VLE platform, but students did not need to submit their own interpretation or self-assessment.

Such an approach controls the frequency of self-assessment, so students would not feel that they need to engage in self-assessment every week from the tenth week on. Second, the class discussion and sharing enabled students to learn from each other and saw what their classmates learnt from professional interpreters. Expert demonstration resources served as a basis to support students' self-assessment as they could compare their interpretation with the experts'.

Students were invited to fill out an end-of-semester survey and participate in focus group interviews on a voluntary basis. In total, 16 undergraduate students and seven graduate students filled out the survey. A total of 22 students took part in the focus group interviews, including 15 undergraduate students and all 6 graduate students.

The survey aimed to understand students' satisfaction rate and their opinions on the effectiveness of the STT as well as the expert demonstration resources. The interviews, however, aimed to check (1) if STT-generated transcripts and expert demonstration resources helped students save time to carry out self-assessment and (2) if these tools and resources helped students gain a better understanding of interpreters' strategies.

12.4   Results and discussions

According to students' responses, over 90% of the students felt that STT was helpful to save the time needed to transcribe their interpretation, as shown in Figure 12.1 and extract.

R: After using Yating?

S7:   It was really fast. However, some words were not accurate, so you need to make corrections, but it would take less than 20 minutes.

R: So it helped you save time?

S7 & S8: Yes. It did (Extract from Focus Group Interview 2).[1]

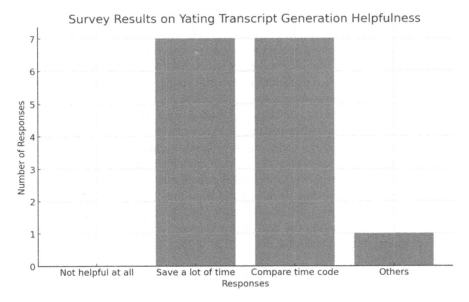

FIGURE 12.1   Survey results on Yating transcript generation helpfulness.

As for the second research question, i.e. when STT technology is used in combination with expert demonstration resources (including audio files and transcripts), to what extent and in which aspects can they help boost student interpreters' confidence when they carry out self-assessment. Most respondents said that the transcript helped them identify their problems related to delivery and fluency, including the problems of unfinished sentences and fillers, as shown in Figure 12.2.

At the focus group interview, students said that expert demonstration resources were really helpful as they can learn different coping strategies after listening to the demonstration.

> S3: I think it is great. Mainly it showed us what to do when you encounter a crisis during the interpreting process or what strategies you can use to gain a few seconds for yourself.
> S2: I think the interpreters were examples we can imitate because there were strategies (from their interpreting). For example, one strategy was to understand first, then interpret. It was really worth learning. And I realized that it was humanly possible. How could they understand those vague words.[2]

The author was also interested to find out students' view about the pedagogical approach and if undergraduate students and graduate students had different opinions. Although both undergraduate and graduate students liked the way the course was conducted, at the interview, they shared different views regarding

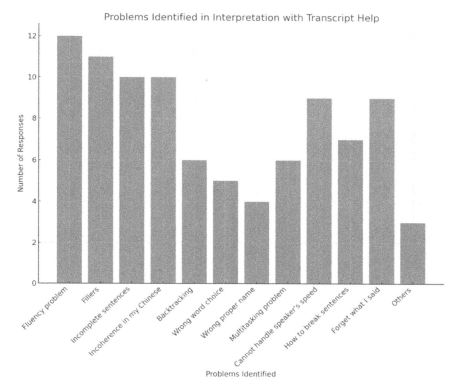

FIGURE 12.2   Problems identified in interpretation with transcript help.

their confidence in future interpreting. For instance, one undergraduate student said:

> I think you can hear how professionals handled those things, but because we would have a comparison version, and you said we should learn their coping strategies and etc., but I felt that…okay… You know how he dealt with (the problem) this time, but two weeks later there was a new speech and (for) the new speech, when I practiced in class for the first time, it felt like it was back to square one. It was the same like before, because the content was different, so the strategies may be different, and I could not cope, so I may feel that I was back to zero.
>
> *(S15)[3]*

However, the graduate students showed more positive attitudes. For instance, one student said:

> But what surprised me the most was that I thought there would be a speech every week, and I might need to find a bunch of PowerPoint materials, but it

turned out that you did not rush us like we were trying to catch a train. You just gave us one speech to practice for 2 to 3 weeks, so we could become familiar with the speaker's speed and his related content, and we were able to, bit by bit, from not complete at the beginning and gradually become more familiar with the speech and we were able to deliver smoother (interpretation).

And the most important thing was to use the skills. At the same time, we could refer to the demonstration of you or those seniors (experts) and learn the translation skills from them, and from you. From this, I can understand that so simultaneous interpreting was when we faced the same paragraph, they would have different strategies and it was worth learning. I think this was the biggest lesson I got from the simultaneous interpreting class.

(M4)[4]

12.5 Conclusion

This study aimed to explore students' opinions regarding the use of STT and expert demonstration resources to support their self-assessment and boost their confidence to engage in self-assessment. The results indicate that most students were satisfied with the use of STT. Although transcripts generated by the STT still showed errors, the use of STT has indeed helped them save time and help them identify their problems more clearly. Second, expert demonstration resources helped students see how experts use strategies interpreting the same speeches and how experts cope with challenges. The interviews also showed that undergraduate students were less confident to use the same strategies for later speeches and graduate students seemed to be more confident. In short, the present study indicates that STT was effective and expert demonstration was effective to support students when they engage in self-study and self-assessment. More detailed analysis of the students' interpretation may help to reveal changes over the eight weeks when they were supported by expert demonstration resources.

Acknowledgment

The project was funded by the MOE Teaching Practice Program.

Notes

1 筆者：現在用了雅婷之後?
S1: 真的很快，只是有時候它可能有些字還是會跑掉，然後你要再花一點時間去修改，但這些前後也不到 20 分鐘。
筆者：所以有幫你們節省時間？
筆者：所以有幫你們節省時間？
S7 & S8: 對，有。(焦點訪談 2)

2 S3：我也覺得這樣很不錯。而且主要是可以讓大家知道說，你如果真的在翻譯當下遇到危機的話有什麼處理方法，或是你也可以用什麼樣的策略替這裡自己爭取多那幾秒。
S2：我覺得口譯員是讓我們可以效仿的對象，因為他裡面有就是一些策略啊，譬如說就是先理解之後再翻譯這種策略，值得我們學習，還可以去了解說原來有人類可以辦到，為甚麼連那種這麼模糊的字都可以聽的出來還翻得出來。

3 我覺得就是可以聽到專業人士是怎麼處理那些東西，但是就是因為我們會有對照的版本嘛，然後老師不是會說要我們學習他的處理方式啊之類的，但是就會覺得說，好，你這次可能知道他怎麼處理了，但是隔兩個禮拜後又是新的一篇，那新的一篇，第一次在課堂上練習的時候就，感覺都打回原形，就是之前的東西，因為又是不一樣的內容，所以可能處理方式不一樣，那我又會沒有辦法去，應付過來就還是就有可能回到零的感覺。(S15)

4 可是其實也讓我最意外的就是 我以為是每個禮拜會有一點點演講丟過來，然後可能要找一堆 PPT 資料，結果發現其實老師的速度沒有想像中的操兵趕馬、起火車這樣，就一個演講讓我們練習個2-3個禮拜，去讓我們熟悉講者的速度以及他的相關內容，能夠一點一滴去⋯可能一開始講的二二六六到慢慢之後越熟悉演講，就越能夠講得更通順。
然後其實最重要的還是運用技巧，同時參考老師或者是學姊們的範例這樣，就是學習他們的，還有跟老師翻譯技巧，這我就可以了解到說原來同口就是在翻同一段的時候，原來他們的策略是不一樣，是值得學習的。我覺得這是在同口中最大的收穫。 (M4)

References

Arumí Ribas, Marta. 2012. "Problems and Strategies in Consecutive Interpreting: A Pilot Study at Two Different Stages of Interpreter Training." *Meta: Translators' Journal* 57(3): 812–35.

Bendazzoli, Claudio, and Annalisa Sandrelli. 2005. "An Approach to Corpus-Based Interpreting Studies: Developing EPIC (European Parliament Interpreting Corpus)." In *MuTra 2005 – Challenges of Multidimensional Translation: Conference Proceedings*. Saarbrücken.

Carabelli, Angela. 1999. "Multimedia Technologies for the Use of Interpreters and Translators." *The Interpreters' Newsletter* 9: 149–55. https://hdl.handle.net/10077/2217.

Cervato, Emanuela, and Donatella de Ferra. 1995. "'Interprit': A Computerised Self-Access Course for Beginners in Interpreting." *Perspectives* 3(2): 191–204. https://doi.org/10.1080/0907676X.1995.9961260.

Chang, Chia-Chien, and Yung-Wei Hao. 2008. "The Creation of an Online Learning Community in Interpreter Training." *Studies of Translation and Interpretation* (11): 119–37. https://doi.org/10.29786/STI.200812.0005.

Class, Barbara, and Barbara Moser-Mercer. 2006. "Designing Learning Activities for Interpreter Trainers: A Socio-Constructivist Approach to Training." In *Proceedings of Society for Information Technology and Teacher Education International Conference 2006*, edited by Caroline M. Crawford, Roger Carlsen, Karen McFerrin, Jerry Price, Roberta Weber, and Dee Anna Willis, 295–300. Waynesville, NC: Association for the Advancement of Computing in Education (AACE).

Class, Barbara, and Barbara Moser-Mercer. 2013. "Training Conference Interpreter Trainers with Technology–A Virtual Reality." In *Quality in Interpreting: Widening the Scope*, edited by Olalla Garcia Becerra, Esperanza Macarena Pradas Macias, and Rafael Barranco-Droege, 293–313. Granada: Editorial Comares.

Corpas Pastor, Gloria. 2018. "Tools for Interpreters: The Challenges That Lie Ahead." *Current Trends in Translation Teaching and Learning E* 5: 138–84.

Defrancq, Bart, and Claudio Fantinuoli. 2021. "Automatic Speech Recognition in the Booth: Assessment of System Performance, Interpreters' Performances and Interactions in the Context of Numbers." *Target* 33(1): 73–102.

Deysel, Elizabeth, and Harold Lesch. 2018. "Experimenting with Computer-Assisted Interpreter Training Tools for the Development of Self-Assessment Skills: National Parliament of RSA." In *Interpreting and Technology*, edited by Claudio Fantinuoli, 61–90. Berlin: Language Science Press.

Díaz-Galaz, Stephanie, Presentacion Padilla, and María Teresa Bajo. 2015. "The Role of Advance Preparation in Simultaneous Interpreting: A Comparison of Professional Interpreters and Interpreting Students." *Interpreting* 17(1): 1–25. https://doi.org/10.1075/intp.17.1.01dia.

Ericsson, K. Anders, and Kyle W. Harwell. 2019. "Deliberate Practice and Proposed Limits on the Effects of Practice on the Acquisition of Expert Performance: Why the Original Definition Matters and Recommendations for Future Research." *Frontiers in Psychology* 10. https://doi.org/10.3389/fpsyg.2019.02396.

Ericsson, K. Anders, Ralf T. Krampe, and Clemens Tesch-Römer. 1993. "The Role of Deliberate Practice in the Acquisition of Expert Performance." *Psychological Review* 100(3): 363-406. https://doi.org/10.1037/0033-295X.100.3.363.

Fantinuoli, Claudio. 2017. "Computer-Assisted Interpreting: Challenges and Future Perspectives." In *Trends in E-Tools and Resources for Translators and Interpreters*, edited by Gloria Corpas Pastor and Isabel Durán-Muñoz, 153–74. Berlin: Brill.

Fantinuoli, Claudio. 2022. "Conference Interpreting and New Technologies." In *The Routledge Handbook of Conference Interpreting,* edited by Michaela Albl-Mikasa and Elisabet Tiselius N, 508–22. London: Routledge.

Fowler, Yvonne. 2007. "Formative Assessment: Using Peer and Self-Assessment in Interpreter Training." In *Critical Link 4: Professionalisation of Interpreting in the Community: Selected Papers from the 4th International Conference on Interpreting in Legal, Health and Social Service Settings*, edited by Cecilia Wadensjö, 253–62. Amsterdam: John Benjamins.

Frittella, Francesca. 2021. "Computer-Assisted Conference Interpreter Training: Limitations and Future Directions." *Journal of Translation Studies* 1(2): 103–42.

Han, Chao. 2018. "A Longitudinal Quantitative Investigation into the Concurrent Validity of Self and Peer Assessment Applied to English-Chinese Bi-Directional Interpretation in an Undergraduate Interpreting Course." *Studies in Educational Evaluation* 58: 187–96. https://doi.org/https://doi.org/10.1016/j.stueduc.2018.01.001.

Han, Chao, and Qin Fan. 2020. "Using Self-Assessment as a Formative Assessment Tool in an English-Chinese Interpreting Course: Student Views and Perceptions of Its Utility." *Perspectives* 28(1): 109–25.

Hartley, Anthony, Ian Mason, Gracie Peng, and Isabelle Perez. 2003. "Peer-and Self-assessment in Conference Interpreter Training." *Pedagogical Research Fund in Languages, Linguistics and Area Studies.* https://researchportal.hw.ac.uk/en/publications/peer-and-self-assessment-in-conference-interpreting-training

Kajzer-Wietrzny, Marta, and Maria Tymczynska. 2014. "Integrating Technology into Interpreter Training Courses: A Blended Learning Approach." *inTRAlinea.* https://www.intralinea.org/archive/article/210.

Lee, Yun-Hyang. 2005. "Self-Assessment as an Autonomous Learning Tool in an Interpretation Classroom." *Meta: Translators' Journal* 50(4). https://www.erudit.org/en/journals/meta/2005-v50-n4-meta1024/019869ar/

Lee, Zi-ying. 2016. "Reflective Practice for Student Interpreters: A Case Study." *Studies of Translation and Interpretation* 2(20): 75–96. https://www.airitilibrary.com/Article/Detail/a0000024-201612-201708240010-201708240010-75-96

Lee, Sang-Bin. 2017. "University Students' Experience of 'Scale-Referenced' Peer Assessment for a Consecutive Interpreting Examination." *Assessment & Evaluation in Higher Education* 42(7): 1015–29. https://doi.org/10.1080/02602938.2016.1223269.

Lee, Jieun. 2018. "Feedback on Feedback: Guiding Student Interpreter Performance." *The International Journal of Translation and Interpreting Research* 10(1): 152–70.

Lee, Juyeon. 2022. "Comparing Student Self-Assessment and Teacher Assessment in Korean-English Consecutive Interpreting: Focus on Fidelity and Language." *INContext: Studies in Translation and Interculturalism* 2(3): 58–83. https://doi.org/10.54754/incontext.v2i3.27

Li, Xiangdong. 2018. "Self-Assessment as 'Assessment as Learning' in Translator and Interpreter Education: Validity and Washback." *The Interpreter and Translator Trainer* 12(1): 48–67. https://doi.org/10.1080/1750399X.2017.1418581.

Li, Tzu-Hsuan. 2019. "Conference Interpreting Preparation: Differences between Expert and Novice Interpreters." Master's thesis, National Taiwan Normal University. Cham, Switzerland.

Liu, Menglian. 2021. "Computer Assisted Student Interpreters' Self-Assessment: Ways and Inspiration." In *Learning Technologies and Systems: 19th International Conference on Web-Based Learning*, 461–71. Switzerland: Springer International Publishing AG.

Mellinger, Christopher D. 2019. "Computer-Assisted Interpreting Technologies and Interpreter Cognition: A Product and Process-Oriented Perspective." *Tradumàtica* 17: 33–44.

Merlini, Raffaella. 1996. "Interprit – Consecutive Interpretation Module." *The Interpreters' Newsletter* 7: 31–41. https://hdl.handle.net/10077/8989.

Monti, Cristina, Claudio Bendazzoli, Annalisa Sandrelli, and Mariachiara Russo. 2005. "Studying Directionality in Simultaneous Interpreting through an Electronic Corpus: EPIC (European Parliament Interpreting Corpus)." *Meta: Translators' Journal* 50(4). https://www.erudit.org/en/journals/meta/2005-v50-n4-meta1024/019850ar/

Motta, Manuela. 2011. "Facilitating the Novice to Expert Transition in Interpreter Training: A 'Deliberate Practice' Framework Proposal." *Studia Universitatis Babes-Bolyai – Philologia* (1): 27–42. https://www.ceeol.com/search/article-detail?id=206270

Ortiz, Luis E., and Patrizia Cavallo. 2018. "Computer-Assisted Interpreting Tools (CAI) and Options for Automation with Automatic Speech Recognition." *Tradterm* 32: 9–31. https://doi.org/10.11606/issn.2317-9511.v32i0p9-31.

Pisani, Elisabetta, and Claudio Fantinuoli. 2021. "Measuring the Impact of Automatic Speech Recognition on Number Rendition in Simultaneous Interpreting." In *Empirical Studies of Translation and Interpreting*, edited by Caiwen Wang and Binghan Zheng, 181–97. New York: Routledge.

Sandrelli, Annalisa. 2005. "Designing CAIT (Computer-Assisted Interpreter Training) Tools: Black Box." In *MuTra–Challenges of Multidimensional Translation: Conference Proceedings*, Saarbrücken 2–6 May 2005, Proceedings edited by Heidrun Gerzymisch-Arbogast (Saarbrücken) and Sandra Nauert (Saarbrücken), 191–209. Advanced Translation Research Center (ATRC), Saarland University, Germany.

Sandrelli, Annalisa, and Jesus De Manuel Jerez. 2007. "The Impact of Information and Communication Technology on Interpreter Training: State-of-the-Art and Future Prospects." *The Interpreter and Translator Trainer* 1(2): 269–303. https://doi.org/10.1080/1750399X.2007.10798761.

Tang, Shang-Yune, and Ming-Li Ju. 2023. "Student Interpreters' Practices and Perceptions of Self-Assessment and Their Implications." *Fu Jen Journal of Foreign Languages* 20: 111–49.

Tiselius, Elisabet. 2013. "Expertise without Deliberate Practice: The Case of Simultaneous Interpreters." *The Interpreters' Newsletter* 18: 1–15.

Tiselius, Elisabet. 2018. "Deliberate Practice: The Unicorn of Interpreting Studies." *Translation–Didaktik–Kompetenz* 131: 131–44.

Yang, Zhimiao, Riccardo Moratto, and Irene A. Zhang. 2023. "China: A Survey on Interpreter Training in China During the Pandemic." In *Educating Community Interpreters and Translators in Unprecedented Times*, edited by Miranda Lai, Oktay Eser and Ineke Crezee, 117–43. Cham: Springer International Publishing AG.

Zhao, Nan. 2022. "Use of Computer-Assisted Interpreting Tools in Conference Interpreting Training and Practice during COVID-19." In *Translation and Interpreting in the Age of COVID-19*, edited by Kanglong Liu and Andrew K. F. Cheung, 331–47. Singapore: Springer.

INDEX

Note: **Bold** page numbers refer to tables.